CARCERAL SPACES

Prisons and immigration detention facilities ostensibly draw sharp divisions between who is inside and who is outside, who is good and who is bad, who is included and who is excluded from society. Contributors to this important volume undermine these dualisms with rich empirical evidence and strong theoretical elaboration that advance the burgeoning field of carceral geography, and offer fresh perspectives to migration studies and criminologists' study of 'punishment and society'.

Jenna Loyd, co-editor of *Beyond Walls and Cages* (2012)

Engaging, thought-provoking and insightful, Carceral Spaces *shines a much-needed light on contemporary practices of incarceration and detention. Required reading for anyone interested in confinement and the control of 'problematic' populations in a globalized world.*

Alexandra Hall, University of York, UK and author of *Borderwatch: Cultures of Immigration, Detention and Control* (2012)

Carceral Spaces
Mobility and Agency in Imprisonment and Migrant Detention

Edited by

DOMINIQUE MORAN
University of Birmingham, UK

NICK GILL
University of Exeter, UK

DEIRDRE CONLON
Saint Peter's University, USA

Routledge
Taylor & Francis Group
LONDON AND NEW YORK

First published 2013 by Ashgate Publishing

2 Park Square, Milton Park, Abingdon, Oxon OX14 4RN
711 Third Avenue, New York, NY 10017, USA

Routledge is an imprint of the Taylor & Francis Group, an informa business

First issued in paperback 2016

British Library Cataloguing in Publication Data
A catalogue record for this book is available from the British Library.

The Library of Congress has cataloged the printed edition as follows:
Carceral spaces : mobility and agency in imprisonment and migrant detention / [edited] by Dominique Moran, Nick Gill and Deirdre Conlon.
pages cm
Includes bibliographical references and index.
ISBN 978-1-4094-4268-4 (hardback)
1. Imprisonment--Social aspects. 2. Migration, Internal--Social aspects. I. Moran, Dominique. II. Gill, Nick. III. Conlon, Deirdre.
HV8705.C37 2013
365–dc23

2012038863

ISBN 978-1-4094-4268-4 (hbk)
ISBN 978-1-138-24934-9 (pbk)

Contents

PART II: SPACE AND AGENCY

List of Figures and Tables

Figures

Table

Notes on Contributors

Deirdre Conlon is Assistant Professor of Sociology and Urban Studies at Saint Peter's University, Jersey City, NJ. Her research is concerned with geopolitics and governance, as well as activism and socio-legal matters in migration and (im) mobility, particularly among asylum seeker and refugee populations. Deirdre is a collaborator in the Asylum-Network research group. Her publications include articles in *Citizenship Studies* (forthcoming), *Population, Space and Place*, and *Environment and Planning D: Society and Space*; she has also contributed book chapters to a number of edited volumes focused on migration and mobility.

Julie de Dardel was born in Geneva, Switzerland. She studied Economic and Social History at the Faculty of Social Sciences of the University of Geneva, with an academic year spent at McGill University in Montreal. Her Master's thesis 'The personal is political': Sexual Revolution and the Women's Liberation Movement received the Ador History Prize in 2006 and was published by Antipodes Editions in 2007. Her PhD research 'International Circulation of the US Prison Model: Impacts in Colombia' at the Geography Institute of the University of Neuchatel is supported by the Swiss National Science Foundation. She is presently a visiting researcher at the Institute of Political Studies and International Relations (IEPRI) of the National University of Colombia in Bogota. Her research interests include social and cultural geography, imprisonment, and international mobility of penal policies, social movements and gender studies.

Nick Gill is Senior Lecturer in human geography at Exeter University, UK. His interests span legal geography, state theory, governmentality and mobilities, with a specific focus on asylum migration, detention and the everyday practices of border control. He is currently leading an ESRC-funded investigation into the legal geographies of the British asylum system and is writing a monograph for Wiley-Blackwell entitled *Peopling Immigration Control: Geographies of Governing and Activism in the British Asylum System*. He has published work in numerous journals including *Transactions of the Institute of British Geographers*, *Progress in Human Geography*, *Antipode* and *Political Geography*.

Nancy Hiemstra is Assistant Professor in the Department of Cultural Analysis and Theory at Stony Brook University in Stony Brook, New York. Her research interests focus on human mobility, migration policy-making, and 'homeland security' at the scales of home and community. Previous research examined Latino immigration to new US destinations. Current research focuses on impacts in Ecuador of migration to the US, as well as migrant detention and deportation policies and practices in the US. She has published articles in *Antipode*, *Gender, Place, and Culture*, *Geopolitics*, and *Social and Cultural Geography*, and has contributed chapters to numerous edited volumes.

Yvonne Jewkes is Professor of Criminology at the University of Leicester, UK. Her main current research interest is the impact of architecture, design and technology on the lives of the prisoners and staff who occupy carceral spaces; and the potential role of computer mediated technologies on the everyday lives and future prospects of prisoners. Yvonne is Founding Editor (with Chris Greer and Jeff Ferrell) of *Crime, Media, Culture: An International Journal* and is also on the Editorial Board of the *British Journal of Criminology*. She is author of the best-selling *Media and Crime*, now in its second edition (Sage, 2004/2011). She has recently guest edited (with Ben Crewe) a special issue of *Punishment & Society* on 'The Pains of Imprisonment Revisited' and a special issue of *Prison Service Journal* on the theme of 'Prison Space'.

Mason McWatters is currently a doctoral candidate in Geography at The University of Texas at Austin. His research broadly considers phenomenologies of confinement, isolation and spatial disorders according to first-hand, testimonial accounts. For his dissertation, Mason explores how proper logics and frameworks of the world disintegrate during crises of agoraphobia, as well as how agoraphobic sufferers struggle to rebuild their worlds in the wake of these 'unworlding' disasters. Theoretically, he is interested in developing a hybrid style of geographical thought that blends materialist and humanist perspectives

Lauren L. Martin is a Postdoctoral Researcher in the Mobilities, Borders, and Identity Research Group in the Geography Department, University of Oulu Finland. While completing her PhD at the University of Kentucky, she was a Visiting Scholar with Grassroots Leadership, with whom she helped found the Hutto Visitation Program in Austin, Texas. She has published on US homeland security knowledge practices, the US-Mexico border wall, and the US detention and deportation system, and her current research focuses on the role of private security companies in immigration enforcement.

Bénédicte Michalon is a Researcher at CNRS (Centre National de la Recherche Scientifique) at Pessac, France. She completed her PhD at 'Migrinter' (Migration internationales, espaces et sociétés) and was a post-doctoral student at the EHESS (L'Écoles des hautes études en sciences sociales), Centre for the Study of Russia, Caucasian and Central-Europe. Her current research focuses on interactions between mobility and international borders, specifically between Romania and the Republic of Moldova, and focuses on the reconstitution of borders in the Eastern enlargement of the European Union, and increased flows of transit and immigration in post-communist countries. She is also working on the role of international migration trends in agricultural areas of southern France.

Olivier Milhaud is a Research and Teaching Fellow in the Department of Geography and Planning, University of Paris Sorbonne, France. He has studied at the Universities of Paris Sorbonne, Lyon, and Bordeaux; in Sweden at Stockholm, and also in the UK at Bristol. His PhD, from the University of Bordeaux, considered prisons as a geographical imagination.

Matthew L. Mitchelson is Assistant Professor of Geography, Kennesaw State University, USA. An economic geographer who studies imprisonment in the United States, his research and teaching interests generally include urban-economic geography, political geography, and GIS. He is particularly interested in multi- and mixed-method research and theories of political economy. His current research includes: a quantitative, cartographic spatial analysis of Georgia prisoners' homes and the prisons in which they were held; a qualitative study of the 'migrations' experienced during imprisonment; and a critical discourse analysis of prison privatization debates. He also co-edited (with Jenna Loyd and Andrew Burridge) the *Beyond Walls and Cages* book project, which seeks to put activists and academics working on prison abolition and (im)migrant justice in conversation with one another.

Dominique Moran is Senior Lecturer in Human and Carceral Geography at the School of Geography, Earth and Environmental Sciences, University of Birmingham, UK. Her ESRC-funded research projects include the first interdisciplinary research, at the interface between geography and criminology/prison sociology, into women's experience of imprisonment in contemporary Russia, and an investigation of the relationship between prison visitation and recidivism in the UK. She is author of the forthcoming book *Carceral Geography: Prisons, Power and Space* and has published articles in journals including *Transactions of the Institute of British Geographers*, *Environment and Planning D: Society and Space*, and *Gender, Place and Culture*.

Alison Mountz Alison Mountz is Associate Professor of Geography and Canada Research Chair in Global Migration. She was the 2009–2010 Mackenzie King Research Fellow with the Canada Program at Harvard University. She is cross-appointed between the Balsillie School of International Affairs and the Department of Geography and Environmental Studies at Wilfrid Laurier University. Dr. Mountz's work explores the tension between the decisions and desires that drive human migration and the policies and practices designed to manage immigration. Her current research examines struggles over border enforcement, asylum, and detention. Her research has been funded by the John D. and Catherine T. McArthur Foundation, the Canadian Embassy, the National Science Foundation, and the Metropolis Project. In 2010 Mountz published *Seeking Asylum: Human*

Smuggling and Bureaucracy at the Border (University of Minnesota Press). The book offers an ethnography of the state that looks at how Canada and other countries respond to human smuggling with changes in border enforcement. *Seeking Asylum* was awarded the 2011 Meridian Book Prize from the Association of American Geographers.

Kelsey Nowakowski studied human geography at the University of Miami, Florida, where she wrote her honours thesis on the topic of inmate labour in the United States and presented it at the 2010 Annual Meeting of the American Association of Geographers. Kelsey is editorial and graphics researcher for *National Geographic* Magazine, and begins her graduate education at the University of Toronto's Geography Department in September 2012.

Judith Pallot is Professor of the Human Geography of Russia, the University of Oxford. Her early work concerned the social history of the Russian peasantry and rural change in post-1991 Russia. She has been PI of two major projects on Russia's carceral geography; one focused on women prisoners' experiences and, the most recent, on prisoners' relatives. She is also interested in the historical geography of the Gulag and is continuing her work on developing the gulagmaps.org website. She is co-author with Laura Piacentini of *Gender, Geography and Punishment: Women's Experiences of Carceral Russia*, published in 2012 by Oxford University Press

Laura Piacentini is Reader in Criminology at Strathclyde Law School and is the first Westerner to have conducted empirical and theoretical work in Russian prisons. For 15 years, she has published widely on the topic of Russian penality, and her work has been used to develop research networks in prison ethnography, research methodologies and transitional justice concepts. She is contributing in a meaningful way to academic debates on the theorisations of the relationship between sociology and the post-Soviet space. Her current research seeks to make empirical research contributions to the on-going cultures of punishment debates.

Jennifer Turner is a PhD Candidate in Human Geography at Aberystwyth University, UK. Her research interests sit broadly at the intersection of cultural and political geography regarding how the contemporary penal system is integrated into British society. Her PhD channels this interest by exploring the transactions between the 'inside' and 'outside' of the prison environment.

Chapter 1
Introduction

Dominique Moran, Deirdre Conlon and Nick Gill

This edited book defines a new field in geographical research, drawing together the work of a new community of scholars and a growing body of work in 'carceral geography', a term we use to describe the geographical engagement with the practices of imprisonment and migrant detention.

In theory these practices are separate, in that 'mainstream' imprisonment incarcerates 'criminals' for custodial sentences imposed by the prevailing legal system, whilst migrant detention confines irregular or non-status migrants pending decisions on admittance or removal. Migration detention is defined as an administrative, non-punitive measure to facilitate expulsion, across the European Union and in the United States (Leerkes and Broeders 2010). However, in practice these spheres increasingly overlap, in complex and intriguing ways, both in terms of the discourses applied to them, the functionality of their institutions, and the experiences of detained individuals. Just as 'mainstream' prison populations have expanded over the past 25 years, there has also been a veritable explosion in the use of detention for irregular migrants. Migrants are increasingly scrutinized as criminals, so much so that scholars and activists now refer to this nexus as 'crimmigration' (Stumpf 2006). Migrant detention centres increasingly resemble prisons in terms of their regimes and security arrangements. Individual migrants are frequently detained in 'mainstream' prisons and subjected to the same 'corrective' regimes as sentenced prisoners, and in both cases citizenship rights are constrained, either through the disenfranchisement of many prisoners, or the disputed identities of some failed asylum seekers in detention (Griffiths 2012). Bosworth (2012) also writes of the 'dual confinement' of foreign national offenders who serve time both in prison and then in migrant detention whilst awaiting deportation.

Despite the similarities between these systems, the distinctions between them have meant that within the academy more broadly, researchers have tended to define themselves by the context in which they work – either imprisonment or migrant detention. Rarely are the two considered in parallel, but as Martin and Mitchelson (2009: 459) point out, both 'hold human beings without consent by other human beings'. We argue here that human geography is well placed to integrate study of the various practices of holding human beings without consent, focusing as it does on the similarities and commonalities between them, through an inherent and integrative focus on the spatial practices of confinement. This book therefore brings together scholars whose work engages practices of imprisonment and migrant detention, and opens a space within geography and related interdisciplinary fields such as critical prison studies, criminology and prison sociology, for conversation and dialogue across these ever more intertwined spheres. In this volume, we seek for the first time to synthesize these related but previously separate strands to create new insights, in dialogue with emergent and topical agendas in geography around mobility, space and agency, and in relation to international policy challenges such as the functionality of imprisonment, and responses to population migration. The broad conceptual focus and wide geographic net is deliberate and consistent with the goals of sparking insight, dialogue and new connections across ordinarily distinct areas.

Imprisonment and Migrant Detention

Changing political, economic and social relations in this globalized era have yielded increased pressure – and propensity, for some – toward greater mobility across state borders. Coinciding with this trend, and contrasting with the free flow of traded goods and consumer services that are frequently connected to neoliberal ideology and practice, migrants meet with ever expanding efforts to immobilize, exclude and excise. There is a wide swath of techniques employed by state governments to these ends including border securitization measures (Hall 2012, Loyd et al. 2012), interdiction and 'push back' policies (Flynn and Cannon 2009), neo-*refoulement* (Hyndman and Mountz 2008), devolution (Coleman 2009, Lahav 1998) and immigrant detention. Each of these practices entails complex questions dealing with issues of space, territory, geo-politics, agency and (im)mobility, and in recent years there has been a burgeoning of scholarly attention to these matters among geographers.

Of particular concern to critical human geographers is migration-related detention, which refers to 'the practice of detaining – typically on administrative as opposed to criminal grounds – asylum seekers and irregular migrants until they can be deported, their identities established, or their claims adjudicated' (Global Detention Project 2007). This practice is now more or less *de rigueur* as part of states' immigration enforcement strategies in the global North. Moreover, as established immigrant receiving states move to externalize immigration enforcement and processing to peripheral and offshore locations, more detention takes place in transit-states and more and more regions have become entangled in a global detention 'enforcement archipelago' (see Mountz 2011). Alongside these trends there has been a massive increase, over a short time span, in numbers of detained immigrants and the US serves as a striking exemplar. As with its prison population, immigrant detention in the US has expanded rapidly and the state now detains the largest number of immigrants on a global scale. In the 10-year period from 1999 to 2009, the yearly rate of immigrant detention more than doubled from 146,760 to 369,483. In 2010, Immigration and Customs Enforcement (ICE) detained approximately 363,000 immigrants. This represents an increase in daily detention rates from under 7,000 individuals in 1994 to roughly 33,000 individuals in 2010 (DHS 2011, Detention Watch Network n.d.). It also means that non-citizens now represent the fastest growing segment of the prison population in the US (No More Deaths n.d.). While overall numbers are lower in other countries, trends in immigrant detention are similar. For example, in the UK, which has the largest detention estate in Europe, the government detained approximately 27,000 individuals in 2011 (Silverman and Hajela 2012), and the detention estate has increased from a daily capacity for 250 individuals in 1993 to 3,500 in 2011.

These developments warrant critical attention by human geographers in their own right, but for scholars of carceral spaces they also highlight the resonances between imprisonment and migrant detention, in at least four ways. First, as noted above, the growth in numbers of individuals being held in immigrant detention facilities parallels increases in mainstream prison populations in the US as well as many other regions since the 1970s (see Aebi and Delgrande 2009). Second, as the privatization of prisons has proceeded, immigrant detention has also become a corporate, for-profit venture. As Flynn and Cannon (2009: 15) observe, 'one of the more notable patterns is how frequently immigration privatization precedes efforts to privatize prisons'. The resonances between prison and immigration industrial complexes (see Fernandez 2007) demand critically informed, theoretically and empirically rich, cross-cutting analyses, such as those offered by contributors to this volume.

Another resonance concerns criminality. For instance, in both the US and the UK, knowingly entering the state using falsified documents now carries both immigration penalties – such as deportation and exclusion from re-entry to the country in the future – and criminal penalties. This is so even when circumstances render these practices unavoidable, such as when individuals flee persecution. The convergence between immigration law and criminal law enforcement does not stop at borders and points of entry however. In the US, once again, over the past two decades 'crimmigration' has expanded and taken root in the interior and has transformed ever greater numbers of irregular immigrants into 'permanent criminals' (Kanstroom 2004, see also Coleman 2007). Immigrants confront the contraction of secure everyday domestic, labour and community spaces, so much so that participation in civil society is fraught with precarity as the threat of being picked up by state authorities – of all stripes – and detained or deported looms as an ever-present possibility. These developments compel scholars to take up questions of space and agency for immigrants as an urgent matter. Contributors to this volume extend this work by examining how detention and spaces in 'civil' society seep into one another (Hiemstra, Martin), by probing how biopolitics and technologies of government operate within detention and prison facilities alike (Gill, de Dardel), and by theorizing possibilities and understandings of autonomy in sites where liberty and agency seem to implode (Conlon, Michalon, McWatters).

With these circumstances, it seems clear that conditions confronted by individuals who are incarcerated within mainstream prisons and detained immigrants overlap in numerous ways. This, then, is a fourth reason for this volume's combined focus from migration and incarceration studies scholars.

Significant areas of overlap include attention to the physical design of prisons (Milhaud and Moran), an issue to which migration support advocates and policy makers have become increasingly attuned. In examining movements into, out of, and around sites of detention (Hiemstra) and incarceration (Moran et al., Mitchelson), authors raise critical policy-oriented questions about the ability to access networks of support as well as offering practical and conceptual insights into the perpetual production of 'ontological insecurity' (Katz 2008), concerns that chime equally in migration and prison studies. Finally, there is much that scholars of carceral spaces can learn from the analysis of popular cultural discourses about prison life that often serve to bolster stereotyped views of incarcerated individuals (Turner). All-in-all, then, these contributions not only advance scholarship in migration and prison studies in their own right, in addition, they propel critical cross-cutting perspectives on all carceral spaces.

Structure

The book emerges out of a series of themed sessions at the conferences of the Association of American Geographers and the Royal Geographical Society with the Institute of British Geographers, and both reflects and extends the current scholarship in the field of carceral geography, and its dialogue and synergy with contemporary critical human geography. The book is arranged in two main sections, themed around Mobility, and Space and Agency, each prefaced by a short introduction by influential scholars in the field – Alison Mountz and Yvonne Jewkes.

Mobility

The first main section, prefaced by Alison Mountz, a leading feminist political geographer who has examined migration and (im)mobility across a wide range of international settings, engages with the notion of mobility, and challenges the assumptions both that confinement necessitates *im*mobility, and that mobility is inherently connected with freedom and autonomy. In so doing, the chapters build upon the recent expansion of literature on the actual or virtual movements of people through space (Cresswell 1999, 2010, Sheller and Urry 2006, Silvey 2004, Urry 2007), including attention to both human and non-human elements (such as waste) and emphasize a deconstructed, post-humanist self, any aspects of which may be mobile or migrate in relative independence (whereas migration scholarship has tended to adopt a more structuralist approach in relation to human agency). The tendency within the mobilities literature has been to understand mobility as a reflection of and a means of reinforcing power (Skeggs 2004); a resource to which not all have equal access, rather than to see it as an instrument of power, which can be used or experienced punitively. Geographers working on imprisonment and migrant detention (e.g. Moran et al. 2012, Gill 2009) have critiqued these understandings of mobility, and point out the forced or punitive mobility which characterizes much of the movement within carceral estates.

In the chapters in this section, these ideas of punitive, governmental or disciplined mobility are taken up and developed further, as authors engage with the mobilities paradigm within a variety of carceral contexts. Nick Gill considers the punitive uses of movement within and outside carceral environments; Bénédicte Michalon analyses mobility and power in relation to migration and asylum policies in the new EU state of Romania; Nancy Hiemstra discusses the 'chaotic' geographies of mobility for US migrant detainees from Ecuador; Matthew Mitchelson engages with imprisonment as a form of migration in relation to the counting of prisoners in the US census; Kelsey Nowakowski considers the intersection between 'immobile' prison labour and the flows of toxic waste through the US penal system; and Dominique Moran, Laura Piacentini and Judith Pallot discuss the forced mobility of prison transport in Russia as a liminal space between 'freedom' and incarceration.

Through this focus on mobility, these chapters make clear the potential for carceral geography to make constructive contributions to contemporary theorizations in critical human geography, and by foregrounding the experience of the confined individual in each of these cases, they open the way for the second section of the book, which takes as its thematic focus agency within carceral space.

Space and Agency

The second section, prefaced by criminologist Yvonne Jewkes whose work is at the forefront of research into prison architecture and design, considers agency within carceral space. Despite an understanding that prison spaces enable actual or perceived constant surveillance, and that this has a direct effect on prisoner behaviour and control (Foucault 1979, Alford 2000), the literature on the built environment of prisons has until recently (Jewkes and Johnston 2007, Fiddler 2010, Hancock and Jewkes 2011) been one of the more underdeveloped areas of penal research (Marshall 2000, Fairweather and McConville 2000). Human geographers interested in secured and carceral spaces recognize space as more than the surface where social practices take place (Gregory and Urry 1985, Lefebvre 1991, Massey 1994); it is understood instead as simultaneously the medium and the outcome not only of political or macro-economic practices, but also of everyday social relations across all spatial scales (Soja 1985). As Adey (2008: 440) argues, 'specific spatial structures... can work to organize affect to have certain effects upon motion and emotion'. Designers of spaces consider 'seductive spatiality' (Rose et al. 2010: 347) or 'ambient power' (Allen 2006: 445) which shape human behaviour within these spaces (Allen 2006, Adey 2008, Thrift 2004). Jones et al. (2010: 492) observe the 'spatialities of behaviour change' in that the design of spaces affects the kinds of decisions that individuals make, both in terms of the macro-scale re-engineering of spatial environments, and the micro-scale recalibration of the fabric of everyday spaces, which 'encourage the normalization of certain patterns of conduct'. In a context in which spaces of confinement are commonly assumed, in the context of Foucauldian panopticism and biopower, to create 'docile' bodies, or Agambenian 'bare life', the chapters in this section of the book engage directly with the notion of agency in carceral space, identifying both ways in which confined individuals are denied rights, and ways in which they contest regimes of confinement and deploy spatial strategies, at different scales and in different settings.

Deirdre Conlon opens this section by discussing the practice of hunger strikes by asylum seekers in Ireland in the context of Foucault's writings on governmentality, arguing that rather than single acts of resistance to 'bare life', they can instead be understood as counter-conduct, a practice that enacts a right to question how subjects are governed. Julie de Dardel considers individual and collective agency of prisoners within a Colombian prison system absorbing a *New Prison Culture* inspired by the US prison system. Olivier Milhaud and Dominique Moran present findings on prisoner agency and privacy in French and Russian prisons, identifying the ways in which prisoners in these contexts consciously negotiate tactical spatial manoeuvres to find solitude in crowded prison spaces. Lauren Martin discusses the complex carceral and legal geographies of immigrant detention in the US, in which noncitizens' rights are limited and the federal government's discretion to detain is extended in and through immigration law. Mason McWatters presents prisoners' poems as a means of understanding prisoner agency within carceral spaces in the US, which he describes as replete with temporal flows and social encounters, and Jennifer Turner discusses the politics of carceral spectacle through the televising of prison life, considering the ways in which media representations of prison life, though opening carceral space to public view, make public prisoners' own stories and experiences.

The final chapter 'Confinement, mobility, space and agency in incarceration and migrant detention' draws together the discourses within and between the foregoing chapters, drawing out the insights and perspectives generated, and identifying the relationship between the theoretically informed and engaged, but grounded, empirical research presented in the volume, and the wider public discourses of penal reform and migrant detention.

References

Adey, P. 2008. Airports, mobility and the affective architecture of affective control. *Geoforum* 39: 438-451.

Aebi, M. and N. Delgrande 2011. Council of Europe Annual Penal Statistics – SPACE I Survey 2009. Strasbourg, pc-cp\space\documents\ pc-cp (2011) 3e. Available at http://www.coe.int/ [accessed 21 July 2012].

Alford, C.F. 2000. What would it matter if everything Foucault said about prison were wrong? Discipline and Punish after twenty years. *Theory and Society* 29: 125-146.

Allen, J. 2006. Ambient power: Berlin's Potsdamer Platz and the seductive logic of public spaces. *Urban Studies* 43: 441-455.

Bosworth, M. 2012. Subjectivity and identity in detention: Punishment and society in a global age. *Theoretical Criminology* 16(2): 123-140.

Coleman, M. 2009. What counts as geopolitics, and where? Devolution and immigrant insecurity after 9/11. *Annals of the Association of American Geographers*. 99(5): 904-913.

Coleman, M. 2007. Immigration geopolitics beyond the Mexico-US border. *Antipode* 39(1): 54-76.

Cresswell, T. 1999. Embodiment, power and the politics of mobility: The case of female tramps and hobos. *Transactions of the Institute of British Geographers* 24: 175-192.

Cresswell, T. 2010. Towards a politics of mobility, *Environment and Planning D: Society and Space* 28: 17-31.

Department of Homeland Security, 2011. Immigration enforcement actions: 2011 annual report. DHS Office of Immigration Statistics. Available at http://www.dhs.gov/index.shtm [accessed 21 July 2012].

Fairweather, L. and S. McConville (eds) 2000. *Prison Architecture: Policy, Design, and Experience*. Oxford: Elsevier.

Fernandez, D. 2007. *Targeted: Homeland Security and the Business of Immigration*. New York: Seven Stories Press.

Fiddler, M. 2010. Four walls and what lies within: the meaning of space and place in prisons. *Prison Service Journal* 187: 3-8.

Flynn, M. and C. Cannon 2009. *The Privatization of Immigration Detention: Towards a Global View*. A Global Detention Project Working Paper. Available at http://www.globaldetentionproject.org [accessed 21 July 2012].

Foucault, M. 1979. *Discipline and Punish*. New York: Vintage.

Gill, N. 2009. Governmental mobility: The power effects of the movement of detained asylum seekers around Britain's detention estate. *Political Geography* 28: 186-196.

Gilmore, R.W. 2007. *Golden Gulag: Prisons, Surplus, Crisis, and Opposition in Globalizing California*. Berkeley: University of California Press.

Global Detention Project, 2007. About the Global Detention Project. Available at http://www.globaldetentionproject.org [accessed 21 July 2012].

Gregory, D. and J. Urry (eds). 1985. *Social Relations and Spatial Structures*. London: Macmillan.

Griffiths, M. 2012. Anonymous aliens? Questions of identification in the detention and deportation of failed asylum seekers *Population, Space and Place* 18(6): 715-727.

Hall, A. 2012. *Border Watch: Cultures of Immigration, Detention and Control*. London: Pluto Press.

Hyndman, J. and A. Mountz 2008. Another brick in the wall? Neo-refoulement and the externalization of asylum by Australia and Europe. *Government and Opposition: An International Journal of Comparative Politics* 43: 249-269.

Jewkes, Y. and H. Johnston 2007. The evolution of prison architecture in Y. Jewkes (ed.) *Handbook on Prisons*, Cullompton, Willan
Jones, R., J. Pykett and M. Whitehead 2010. Governing temptation: Changing behaviour in an age of libertarian paternalism. *Progress in Human Geography* 35(4): 483-501
Kanstroom, D. 2004. Criminalizing the undocumented: Ironic boundaries of the post-September 11 'Pale of Law'. *North Carolina Journal of International Law and Commercial Regulation*, 29, 639-670.
Katz, C. 2008. Childhood as spectacle: Relays of anxiety and the reconfiguration of the child. *Cultural Geographies*, 15(1), 5-17.
Lahav, G. 1998. Immigration and the State: The Devolution and Privatisation of Immigration Control in the EU. *Journal of Ethnic and Migration Studies* 24 675-694.
Leerkes, A. and D. Broeders 2010. A Case of Mixed Motives? Formal and Informal Functions of Administrative Immigration Detention *British Journal of Criminology*50 (5): 830-850.
Lefèbvre, H. 1991. *The production of space*. Oxford: Blackwell.
Loyd, J., M. Mitchelson and A. Burridge 2012. *Beyond Walls and Cages: Prisons, Borders, and Global Crisi*s. Athens, GA: University of Georgia Press.
Marshall, S. 2000. Designing control and controlling for design: Towards a prison plan classification for England and Wales?, in J.R. Gold and G. Revill (eds). *Landscapes of Defence*. Harlow: Prentice Hall, 227-245.
Martin, L.L. and M.L. Mitchelson 2009 Geographies of Detention and Imprisonment: Interrogating Spatial Practices of Confinement, Discipline, Law, and State Power. *Geography Compass* 3(1): 459-477.
Massey, D. 1994. *Space, Place and Gender*. Cambridge: Polity Press.
Moran, D., L. Piacentini and J. Pallot 2012. Disciplined mobility and carceral geography: prisoner transport in Russia. *Transactions of the Institute of British Geographers* 37(3): 446-460.
Mountz, A. 2011. Islands as enforcement archipelago: Haunting, sovereignty, and asylum. *Political Geography* 30: 118-128.
No More Deaths n.d. *Prisons and detention.* Available at http://www.nomoredeaths.org [accessed 21 July 2012].
Rose, G., M. Degen and B. Basdas 2010. More on 'big things': building events and feelings. *Transactions of the Institute of British Geographers* 35: 334–349.
Sheller, M. and J. Urry 2006. The new mobilities paradigm. *Environment and Planning A* 38: 207-226.
Silverman, S. and R. Hajela 2012. *Immigration detention in the UK.* The Migration Observatory. Available at www.migrationobservatory.ox.ac.uk [accessed 21 July 2012].
Silvey, R. 2004. Power, difference and mobility: feminist advances in migration studies. *Progress in Human Geography* 28: 490-506.
Skeggs, B. 2004. *Class, Self, Culture*. London: Routledge.

Soja, E. 1985. Regions in context: spatiality, periodicity, and the historical geography of the regional question. *Environment and Planning D: Society and Space* 3: 175-190.

Stumpf, J. 2006. The crimmigration crisis: immigrants, crime, and sovereign power. *Berkeley Electronic Press (bepress) Legal Series* paper 1635. Available at http://law.bepress.com/expresso/eps/1635 [accessed 21 July 2012].

Thrift, N. 2004. Intensities of feeling: towards a spatial politics of affect. *Geografiska Annaler B* 86B: 57-78.

Urry, J. 2007. *Mobilities*. Cambridge: Polity.

PART I
Mobility

Chapter 2
On Mobilities and Migrations

Alison Mountz

This lively and important section of the collection offers reflections on migration and mobility that especially bring to the fore the various kinds of forced mobility that accompany confinement. Considered as a collective, the authors destabilize easy categorizations of people as either mobile or incarcerated subjects. The authors advance what – up until now – have been important, yet preliminary debates about the relationship between forced mobility and the burgeoning literature on mobilities (e.g. Gill et al. 2011). Indeed, careful geographical consideration of the paradoxical forces of migration and containment provides broad, varied and sophisticated insights into the complex relationships between mobility, confinement and freedom. In my short intervention, I consider each author's piece briefly, and then outline key contributions collectively offered and questions collectively opened.

Nick Gill opens the section by offering new insights into the relationship between mobility and liberty. He achieves these insights by carefully reading interdisciplinary literatures on mobility, detention and penology through one another. Gill challenges simplistic correlations and disavowals that have been drawn between mobility and liberty. He questions the tendency among scholars to posit restricted mobility as restricted liberty. Instead, he offers a more sophisticated analysis that understands mobilities and liberty to be entwined in more complex fashion, with mobilities operating as 'vehicle' to alternatively deny or deliver punishment *vis-à-vis* liberty. Gill develops this argument through discussion of electronic monitoring systems used to track former inmates and detainees beyond the spaces of prison. He finds the imposition and maintenance of physical and social mobility to be in itself a punitive measure used to curtail personal liberty. This measure relies on the idea of liberty as a desirable commodity manipulated to create Foucault's classic, self-disciplining subjects of governmentality, with mobility 'as a way to rule *through* the [promise of] freedom of subjects'. Gill understands 'punitive mobility' to operate as a continuous spatiotemporal logic drawn from logics at work *within* prisons. As such, his piece parallels earlier arguments made by migration scholars that illicit border-crossings and punitive immigration policies create forms of self-policing, containment and criminalization where no detention walls exist (e.g. Kearney 1991, Coutin 2010). Gill's essay also sets the stage for the pieces to come, for the notion of 'punitive mobility' and the theme of carceral logics extending beyond the space of the prison recurred through this collection of essays.

If Gill posed demanding questions about the relationship between mobility and liberty, Bénédicte Michalon takes up the issue and offers a response of sorts in her examination of forced mobility in institutional settings as a form of governmentality. Michalon's version of punitive mobility manifests in her analysis of the transfer of problematic detainees. Her methodology involved ethnographic research in Romania in 2009 and 2010. Although the technology of her focus is distinct from Gill's, Michalon's tracing of internal movement through the detention system in Romania maps similar dynamics with the transgression of interior and exterior spaces of the prison. Michalon also builds on scholarship that finds mobility to be a central part of imprisonment. Like Gill, she posits mobility as a tool deployed by the state to exert control. Also like Gill, she finds a continuous set of spatio-temporal punitive logics at work within and beyond spaces of detention. Michalon traces the seemingly contradictory processes of movement and bordering, from careful detailing of the controlling built architecture of Romania's two detention centres to research participants' narratives of controlled circulation. She discovers a paradoxical spatio-temporal logic of confinement and control: the same architecture that inhibits movement also forces movement in controlled fashion, both within and beyond the spaces of detention. Beyond Romanian facilities, this paradox operates in multi-scalar ways as detainees are shuttled among and then beyond the Romanian centres to other facilities across the European Union.

Whereas Michalon finds movements within and among detention facilities to reflect paradoxical forces of confinement and control, Nancy Hiemstra characterizes the spatio-temporal logic of Ecuadorian migrants' movements through the United States' detention and deportation system as 'chaotic.' Building on Weizman's (2007: 8) suggestion that 'Chaos has its particular structural advantages'; Hiemstra sets out to map the work done by the 'immigration industrial complex' through the seemingly chaotic geographies of detention. Like Michalon, Hiemstra draws on painstakingly careful empirical research in the United States and Ecuador among former detainees, deportees and their families. She finds that chaos is not simply an innocent outgrowth of the expansion of detention and imprisonment, but a driving force that arrests the power of forced mobility to control. Like Michalon, Hiemstra focuses on the frequent transfers that are a central component of the US 'D&D' system. She follows the routes and consequences of these transfers, detailing experiences of confusion and isolation as detainees literally and figuratively circle the country in vertiginous fashion. The theme of punitive mobility appears again in Hiemstra's chapter, this time with discussion of the profound psychological and economic damages of waiting as one wades through chaos.

Matthew Mitchelson sustains the discussion of paradoxical forces of fixity and mobility at work in the logics of incarceration, but he adds distinct material from the other authors in this section in his focus on the issue of imprisonment and census-taking in the United States. Mitchelson asks a classic question posed by social scientists: Who counts? He also asks a classic question posed by geographers: where is home? Mitchelson's essay brings together issues of voting, citizenship, participation and the landscapes of power that undergird and undermine the supposed ideas and ideals of democracy. Mitchelson brings into contrast two perspectives on the fragmentation and paradoxical forces driving prison landscapes: those of prisoner and state. By conducting interviews with 50 people who experienced the prison system in the state of Georgia, Mitchelson aims methodologically to challenge the absence of prisoners from recent controversy about their simultaneous inclusion (through counting) and exclusion (denied the right to vote) from representative democracy. His participants' narratives illuminate a fourth expression of 'punitive mobility' across the carceral archipelago: 'incarceration-as-migration'. Here, Mitchelson's former prisoners' experiences of what he calls 'population churning' correspond with the frequent transfers of Hiemstra's former detainees, with enforced movement resonating across regional, national and transnational carceral landscapes.

Kelsey Nowakowski's thoughtful contribution to this collection addresses the exploitation of inmate labour through their simultaneous exclusion from political and social dimensions of society and integration into regional and global economies. She first offers a brief history of the privatization of prison labour in the United States and then launches into a case study of the use of inmate labour in recycling e-waste. Nowakowski looks through the lens of capital accumulation to see the circuits of commodity chain production through which private corporations stand to benefit from the labour inmates do in the realm of recycling e-waste. Building on scholarship that characterizes prisons as 'landscapes of defence' designed to contain hazard and risk, Nowakowski juxtaposes the mobility of the products of migrant labour with the spatial isolation and social marginalization of prisoners themselves. Prisoners-as-workers are alienated from Occupational Safety and Health Administration (OSHA) protections of safe work standards, labour law and legal avenues to challenge labour practices. This alienation means that they bear the brunt of toxic materials on their bodies. Nowakowski maps intimate landscapes of mobility and confinement, with 'punitive mobility' here taking unique expression of the movement of toxic waste onto and into the imprisoned body.

In the final piece, Dominique Moran, Laura Piacentini and Judith Pallot analyse transfer of women prisoners in the Russian Federation and in so doing, challenge easy divisions between carceral spaces inside and outside of the prison. As such, their piece affirms and enhances other contributors' mappings of the movement of spatio-temporal logics between supposedly interior and exterior zones. Moran et al lay the foundation for their study by building on critiques of Erving Goffman's idea of prison as 'total institution.' Rejection of total enclosure enables them to conceptualize the more spatially nuanced 'transcarceral space', a liminal zone that involves circular movement through the confines of prison walls. True to their conceptualization of imprisonment as a transborder phenomenon, Moran et al's collaborative methodology also involved research within and beyond the prisons with current and formerly incarcerated. This methodology makes possible a sustained analysis of liminality that challenges and should enhance existing scholarship on carceral landscapes. Much recent scholarship, inspired often by Giorgio Agamben's ideas, has dwelled in the liminality of border zones. Moran et al show that these forces operate equally intensively in the liminal zones surrounding prisons. Important things transpire in these liminal zones: disappearance, transgression, human rights violations and prisoner protest. Women prisoners in Moran et al's research narrated intense experiences of transcarceral spaces. Indeed, theirs is the only piece to focus on the experiences of women who are often a minority numerically in the prison populations under study, but a group with distinct and chilling tales to tell about punitive mobility.

The major collective contribution made by these authors is the poking, prodding and testing to advance understandings of the complex relationships between mobility, liberty andconfinement. While these relationships are highly varied and dynamic, certain key themes recur across the pieces and demand discussion here.

One recurring theme operating across the essays is state use of mobility as a tool to punish, control and discipline people who are physically confined as well as those who are not. Indeed, time and again we see the movement of the very tactics and logics of punishment and control migrating back and forth between interior spaces of prison and spaces seemingly exterior to facilities. For Michalon, for example, the carceral logics extend beyond the facilities in Romania to other border-crossings and detention facilities across Europe. The extension of carceral spaces and logics beyond the space of prison parallels arguments about the work of criminalization that illicit border-crossings do well beyond the spaces of borders in creating pliable, vulnerable work forces and people without access to citizenship for whom fear shapes many dimensions of daily life. Hiemstra's work brings these two movements beyond the detention facility and beyond the illicit-border crossing together. She finds that the chaos of the US D&D system follows Ecuadorians and their families into the intimate spaces of home in Ecuador. In a reversal of sorts, Nowakowski's mobility of toxic e-waste into prison labour enables the confinement of safety and security beyond the spaces of prison. Highly mobile themselves, carceral impulses to confine become an absent presence, at times hidden, but continuously reappearing, whether in toxic work conditions where mobile toxicity embeds the body in the global economy, or in the criminalizing aftermath of a chaotic journey through various national and transnational detention regimes.

These authors offer their readers insights into the conduct of challenging research by accessing sites and people who have been intentionally hidden by the state. It is noteworthy, exciting and relatively new that so much of this scholarship has been taken up by geographers who collectively approach carceral landscapes as multi-scalar, cross-border, transnational, regionally-integrated and highly localized. Geographers have played and might sustain an important role in emerging subfields including carceral geographies (Moran et al. 2012) and geographies of detention and imprisonment (Martin and Mitchelson 2009, Mountz et al. forthcoming). Furthermore, the geographic breadth of this research is at once exciting, impressive, illuminating and terrifying, for there is much more to be made – as the editors of this collection well know – of the connections across national settings that too often divide anti-prison political struggle.

Collectively, these pieces open a number of questions ripe for further research. Nowakowski's piece, for example, brings the political economy of detention and incarceration to the foreground by looking at the how people are integrated in regional and global economies through their migration, detention and labour. Her work might prompt consideration among the other contributors to re-imagine their analyses through the lens of detainees and prisoners as workers in broader economies.

The pieces also collectively pose a series of challenges around language that pertain at once to nomenclature, conceptual and political issues. In my discussion of each contribution, I have tended to use the same language used by each author in their writing. These have varied greatly, conceptually (liberty, mobility, migration, incarceration, confinement, containment, imprisonment) and in terms of categories of naming (prisoners, inmates, incarcerated, prisons, detention, workers and migrants). These decisions around language are never innocent, of course, but always politicized as they open up chances for solidarity and the drawing of connections across disparate and dispersed experiences of forced mobility.

Collectively, the pieces also invite further inquiry into how detainees, former detainees and prisoners themselves understand their own mobility. Michalon, Hiemstra, Mitchelson and Moran et al each illustrate the importance of the dialogic process wherein state actors and confined individuals narrate distinct perspectives on issues of migration and mobility.

To conclude, this collection is satisfyingly replete with paradox and juxtaposition of the ironies of mobility and migration brought handily to the fore by the authors. Their work demands that questions of social justice come more fully and explicitly into the fold. What can be done with this rich and illuminating material that will engage and enhance anti-prison movements currently underway? What spaces of political engagement have been opened by these authors, in their individual research programs and by the careful assembling of their pieces in this collection? And finally, what forms of transnational engagement and solidarity can the connections that they have mapped activate?

References

Coutin, S. 2010. Confined within: national territories as zones of confinement. *Political Geography* 29(4): 200-208.

Gill, N., J. Caletrío and V. Mason 2011. Introduction: Mobilities and forced migration. *Mobilities* 6(3): 301-316.

Kearney, M. 1991. Borders and boundaries of state and self at the end of empire. *Journal of Historical Sociology* 5: 52-74.

Martin, L. and M. Mitchelson 2009. Geographies of detention and imprisonment: Interrogating spatial practices of confinement, discipline, law, and state power. *Geography Compass* 3(1): 459-477.

Moran, D., L. Piacentini and J. Pallot 2012. Disciplined mobility and carceral geography: prisoner transport in Russia. *Transactions of the Institute of British Geographers* 37(3): 446-460.

Mountz, A., K. Coddington, R.T. Catania and J. Loyd (forthcoming). Conceptualizing detention: mobility, containment, bordering, and exclusion. *Progress in Human Geography.*

Weizman, E. 2007. *Hollow land: Israel's architecture of occupation.* New York: Verso.

Chapter 3
Mobility versus Liberty? The Punitive Uses of Movement Within and Outside Carceral Environments

Nick Gill

Introduction

Many Western democracies have seen increasing imprisonment rates in recent years. Between 2.3m and 2.4m Americans are now behind bars, for example: roughly one in every 100 adults. If those on parole or probation are also included, one adult in 31 is under 'correctional' supervision (*The Economist* 24 July 2010: 23). Similarly in England and Wales, the prison population has risen steadily, from 45,500 in 1992 to 81,833 in November 2007 (Ministry of Justice 2008). In general, the World Prison Population List finds that, of 218 independent countries, 71 per cent are experiencing increasing prison populations (Walmsley 2009). Tonry (2001) traces this rise to, among other things, a phenomenon that he calls 'populist punitiveness': the temptation to curry political favour with electorates by punishing criminals ever-more harshly. '[N]o important constituency is offended when campaigns are launched against welfare recipients, immigrants or criminals, and criminals least of all', he suggests (Tonry 2001: 524).

This political penchant for punitiveness has given rise to an increase in the use of electronic monitoring (EM) as a cost effective way to respond to the increasing number of convicts. Yeh (2010: 1090) reports that electronic tags are issued with increasing frequency for the purposes of tracking offenders in the United States, especially in the case of so-called 'sexual predators'. Paterson (2007), writing about the use of electronic tagging in the UK context, observes that an ever-wider cross-section of the population are also subject to tags, including bailees, adult offenders, juvenile offenders, terrorist suspects and those subject to immigration controls.

This trend can be set within a broader societal securitization that has witnessed the militarization of everyday life (Feldman 2002). The continuing subordination of geopolitics and geo-economics to security issues has accompanied the persistent justification of intrusive and expensive geo-surveillance procedures on the basis of imagined security threats (Graham and Wood 2003, Fisher and Dobson 2003). This fetishization of security manifests in the particular case of EM through the multitudinous suggestions of how to increase its coverage, often for 'good', 'productive' and 'sensible' reasons (Rancière 2004): for example to trace sufferers of Alzheimer's disease who are likely to get lost, to impose a night-time curfew on unruly children, to protect children in shopping centres from potential abductors, to deter stalkers and violent partners from approaching their victims and to improve operational efficiency in the workplace by aiding employee distribution in a range of contexts (see Bloomfield 2001).

It would be understandable to associate the rise in prison numbers in Western democratic countries with the mass confiscation of mobility. Mobility in this chapter is taken to refer to the movement of both human and non-human actors from one place to another at a variety of different scales. In the case of the former this includes movement of asylum seekers, refugees and the homeless; business travellers; students, au pairs and other young people; people travelling for medical purposes; military movement of personnel; post-employment travel; the 'trailing travel' of children, partners and servants (Kofman, 2004); travel of service workers; tourist travel; travel to visit friends and relatives and work-related travel including commuting (Urry 2007: 10-11). This cacophony of different forms and scales of movement especially characterizes modern elite and middle-class Western life, and presents vested interest groups with perceived governance challenges. While access to these different forms of mobility or the means to partake in them is certainly not ubiquitous, the confiscation of mobility *now*, in particular, might therefore be seen as an attack upon an increasingly important way in which individuals are able to take part in modern society.

Prison thus appears to be a way in which the state makes convicts immobile while EM can be seen to limit permitted movements. To the extent that 'mobility is one of the aspects of freedom' (Braidotti 1994: 256), and 'imprisonment necessarily entails depriving an inmate of some degree of personal liberty' (Roberts and Gabor 2004: 95), we might understand incarceration as the withholding of an individual's mobility in order to punish that individual. Indeed in their superb review of geographically informed research into detention and imprisonment, Martin and Mitchelson (2009) define imprisonment and detention in terms of the restrictions placed over 'individuals' ability to move from one place to another' and characterize them as spaces within which 'individual mobility is highly constrained, if not eliminated' (p460; see also Michalon, this volume).

This chapter argues, however, that the eliding of confinement and stasis overlooks both the fact that mobilities are employed within prisons in order to deliver their punitive functions and the ways in which punitive confinement in a broader sense is achieved on the move through technologies such as EM. Rather than seeing EM as a soft option or as a break from the traditional carceral estate, the chapter emphasizes the continuities between traditional prison and EM by seeing EM as a more sophisticated example of the way in which mobility is perfectly commensurate with confinement and has been used as a constituent element of confinement within prisons for many years.

Accordingly, the chapter brings together a variety of disparate literatures in three new ways. First, the chapter demonstrates the ways in which various

sociological and penological theories have strong potential for synergy. Sociologists have recognized the varieties of presence and co-presence that it is possible to experience (Simmel et al. 1997), while writers in penology have drawn attention to the varied ways in which confinement can be achieved (Carnochan 1998). To the extent that forms of confinement correspond to various forms of presence, these disciplines are ripe for a renewed engagement. This engagement might examine how presences and their mechanisms of comportment – mobilities (see Urry 2007) – are becoming reconfigured in ways that are commensurate with the 'revised ambitions of political government' and specifically 'the aspiration to govern 'at a distance'' (Rose 2000: 337). Second, the chapter intentionally fuses work on prisons and work on immigration detention, in the spirit of this edited volume, in order to represent their increasing connectedness as a result of the 'criminalizing discourses surrounding migration' that have justified a range of innovations in the management of migration that draw upon legal and penological systems for inspiration (Bosworth and Guild 2008: 707, see also Webber and Bowling 2008).

Third, the chapter also builds upon a gathering scepticism among scholars of mobilities about the association between mobility and freedom (Sheller 2008, Urry 2007, Cresswell 2010), by drawing upon penological literature both within and outside prisons to highlight a series of oppositions between the two phenomena. By making this argument the chapter goes beyond the position of these authors. While they have recognized that mobility and liberty are not necessarily aligned, this remains too weak a position to appreciate the various ways in which they are directly opposed. The chapter consequently argues that mobilities scholars, penologists and geographers working in the area of detention and confinement have not done enough to recognize the ways in which mobilities can be i) antithetical to liberty or ii) a vehicle through which 'liberty' itself can be used to govern and/or punish (see Rose 1999). In both senses the chapter demonstrates how mobilities are systematically used by states in order to deliver punishments.

This point is illustrated in the case of prisons and detention centres by examining the ways in which inmate mobility forms part of their punitiveness. It transpires that mobility is used by authorities to increase the discomfort, invisibility and insecurity of carceral spaces. Moreover, the chapter also argues that the increasing use of EM demonstrates the ways in which the punitiveness of confinement absolutely does not rely upon immobility. Indeed, acute pains of imprisonment are achieved when the prisoner is afforded a degree of mobility that acts to extend the space of confinement understood as 'restrictions over body and mind' (Carnochan 1998: 380). The case of EM also demonstrates the way mobility can punish through the responsibilization of the convict in ways that are qualitatively distinct from traditional prison spaces.

In conclusion to the chapter, and following from the preceding discussion, a further consideration emerges around whether punitive mobility is necessarily imposed, or whether it can sometimes be volitional (and yet still punitive). While this may sound paradoxical, it is perfectly possible under a governmental

framework for the punished to desire and even execute the punishment, and the concern here is with the spatial manifestation of this government of self by self. Indeed it might be seen as an achievement of a fully-developed and sophisticated penal order, located within a liberal society that demands of its citizens independent and autonomous habits of self-management (Burchell 1996), that it calls forth a penitent prisoner, anxious for absolution and for the chance to be reinserted into productive social circuits (see Conlon and Gill forthcoming). For Cruikshank (1999) this is the apex of a governmental apparatus, one that imbues subjects with the will to take on their own responsibilities, to empower themselves. As the argument of the chapter progresses then, the nature of the relationship between mobility and punition can be seen to alter: from an imposed, blunt sense of movement as discipline and punishment to a more complex situation in which the circulation of prisoners within prisons might conceivably be justified as 'good' for prisoners in terms of their capacity to socialize and to operate normally in a liberal order (see Milhaud and Moran this volume), all the way to electronic monitoring that *at once* permits and sentences the convict to take up their social responsibility and continue to fulfil their work-related and familial obligations. Even as the degree to which the relation between punishment and movement may vary in terms of volition, however, they each encourage us to think critically about the relationship between mobility and liberty. This chapter attends in particular to the circumstances in which the former is actually being used to undermine the latter, either with or without the assent of the governed.

Mobility and Liberty: An Uneasy Relationship

The association between freedom and mobility is common. As Sheller (2008: 25) notes, 'It has often been assumed that mobility equals freedom and that freedom requires mobility'. Deleuze and Guattari (1986: 443), for example, contrast 'free action, related to pure mobility' with 'the order of work with its conditions of gravity, resistance and expenditure'. From a different perspective, Bauman (2000), draws a parallel between mobility and liberty by suggesting that techniques of escape, slippage, elision and avoidance are the key bases of power in modern society (see also Hirschmann 1970). 'The might of the global elite' Bauman writes (2000: 188) 'rests on its ability to escape local commitments'. This renders the powerful increasingly 'mobile, shifty, evasive, slippery and fugitive' (Bauman 2000: 14) while it is the powerless – the un-free – who have to endure stasis and fixity. Such a view also chimes closely with calls to recognize mobility as a human right, as fundamental to human happiness as freedom from the fear of persecution or hunger (Smith 2004).

Urry (2007), however, is cautious about framing mobility as an object of rights, preferring to nuance his understanding of the relationship between mobility and freedom with an appreciation of capabilities (Nussbaum 2006). It only makes sense to see mobility as a right if individuals or groups have the capacity to make

the most of this right. Indeed, in the extreme, if individuals had an impoverished capability for mobility, but were nonetheless expected or forced to be mobile, we can see how mobility might become a form of imposition and injustice. This recalls Sheller's (2008) discussion of 'mobility injustices' – mobility can constitute injustice if it demands more from an individual than their capabilities will allow.

Mobilities studies have also viewed *immobilities* such as stasis and slowness suspiciously. They have typically been seen as tools of entrenched political interests, deployed because political elites see free movement as a 'threat, a disorder in the system, a thing to control' (Cresswell 2006: 26). One of the risks of viewing immobility sceptically, however, is to draw a consequent association between mobility and freedom from domination. Cresswell (2006: 55), for example, warns against the 'romanticization of the nomad' (2006: 55), 'deep-rooted ideas about mobility as a progressive force' (2006: 56) and the pitfalls of simplistic 'mobility-as-liberty' (Cresswell 2010: 27) ways of thinking that are deeply ingrained in everyday understandings of mobility itself. Similarly, Adey (2010: 87) argues that 'The ideological associations of mobility with liberty, freedom and universalism … contain serious shortcomings'.

Empirically, both Sennett (1998) and Favell (2008) have recognized the psychological and emotional burdens that frequent movement can impose. Both authors consider highly-skilled, middle class workers in Western societies, notable for their mobility. While privileged in many ways, both authors conclude that mobility itself constitutes a significant burden to this section of the population. In his study of highly skilled European telecommunications workers, for example, Favell discusses the unpleasant 'weightlessness' that these workers often experience as a result of their constant migrations, both short and long term. Sennett (1998) similarly diagnoses the 'drift' that accompanies modern mobility, which entails ambiguous identities, the corrosion of character and disattachment from community life. In this vein, Favell concludes that 'The social theorists of mobility have it wrong. There are human limits to flexibility, movement, and transience … Ultra-mobility is not a stable long-term option for real people' (Favell 2008: 113).

Welcome though these observations are, however, the debate about the relationship between liberty and mobility seems to largely revolve around discrediting their correlation, which falls far short of theorizing the situations in which they are actively opposed. Only a few studies have dealt with the subjugating nature of many everyday mobilities. In discussing US airspace, for example, Kaplan (2006: 396) conceptualizes some mobilities as 'the suppressed articulation of constraint'. Elsewhere, Cresswell's (2010: 27) call for a politicization of mobilities recognizes the way feudal power in Europe fused exploitation with frequent, compulsory re-location.

In an unpublished work, Elizabeth Shove (2002) comes closest to developing a theory of mobility versus liberty. Considering, again, the case of highly-skilled, middle class workers, she describes the 'mobility burden' that is borne by hypermobile (see Adams 1999) employees who are perpetually 'rushing around'

as a result of the quickening of professional life. This chapter adopts Shove's reservations about mobility in a new context. Thinking beyond the middle classes, it advocates a more general awareness of the ways in which mobility can subjugate different groups. Building on Shove's work, we can begin to conceptualize extreme burdens of mobility that result from the calculated, punitive uses of human movement.

Extreme Mobility Burdens

A body of penological literature examines the political uses, and experiences, of prisoner mobility to, between and within carceral establishments and it is possible to distil various aspects of the association between prisoner mobility and punitiveness from this literature. For the purposes of this chapter the focus is upon three ways in which mobility is used punitively within prison and detention centre space: as a source of discomfort, visibility and exposure.

Inmate mobility may simply be uncomfortable, involving arduous journeys as well as loss of familiarity with surroundings, other inmates and localized sources of support such as families and lawyers. It is noteworthy that 'transportation' was a punishment in itself in much of Europe during the 18th Century, with the hard-working American colonies and Australia the most frequent destinations (Ekirch 1985, Woodward 2006). Describing the birth of the French transportation system in 1719, for example, Hardy (1966: 214) outlines the 'hardship of a journey across northern France in the dead of winter' that the first transportees experienced, exposing them to 'bitter cold, starvation, and disease' (1966: 214). While transportation is no longer in widespread usage as a punishment in itself, there is still concern that the deportation of immigrants, and especially so-called 'illegal' immigrants, continues to be used punitively (Schuster 2005). The conditions of immigration detainee transportation in the UK have prompted investigation for example, demonstrating that the transportation of incarcerated individuals constitutes moments of acute stress for prisoners, especially if they are unsure of their destinations or the purpose of their relocations (HM Inspectorate of Prisons 2006, Medical Foundation of the Care of Victims of Torture (MFCVT) 2004).

Mobility can also reduce prisoners' visibility. In their description of prison conditions in the United Kingdom in the early 1990s, Collins and Burns (1992: 16) outline a process that they refer to as 'ghosting', whereby prisoners who are considered disruptive are relocated from one prison to another in order to disorientate and displace them.

> The practice of 'ghosting' disruptive prisoners, by moving them frequently through the prison system, often putting them in solitary confinement, is one way the prison system deals with inmates who are 'control problems' … 'Ghosting' has been condemned by prison monitoring groups as an intimidation tactic …

> When relatives attempt to visit the prisoners, they arrive at the prisons only to be told that the prisoner is no longer there.

The movement of prisoners in this way can amount to not just a one-off action but to an institutional tactic. Feldman (1991) discusses the unannounced segregation of prison groups and swapping of prisoners between cells as a tactic to disorientate individual prisoners and to weaken the 'dirty' protest that was mounted by Irish Republican prisoners in Her Majesty's Maze Prison near Belfast in Northern Ireland in the 1970s and 80s. Prisoners describe their concern that they were incarcerated outside the purview of both the larger prisoner population within the prison and the political community of which they were a part outside the prison walls. This, they felt, exposed them to ill-treatment that often went un-noticed. 'Nobody knows we're here' one inmate recounts 'nobody knows what's going on' (Feldman 1991: 159).

There are concerns that similar uses of mobility, that render the incarcerated invisible, are employed today and demonstrate the subtle links between geopolitical systems and techniques of securitization, confinement and control (Rose 2000). Sparke, for example, outlines the nature of extraordinary rendition used by the US as 'another penal development using subcontracting, but one that deliberately outsources overseas' (Sparke 2006: 173). This has allegedly given rise to 'ghost flights' that transport terror suspects into extraordinary spaces characterized by legal uncertainty (Doward 2009). Mountz (2011a) uses the related metaphor of haunting to describe the ways in which detainees are subjected to seemingly unaccountable state power that depersonalizes migrants by locating them in remote carceral establishments and ambiguous legal situations.

The transportation of prisoners can also expose inmates to situations in which they feel that their personal security is threatened. Spierenburg (1996: 35), for example, emphasizes the punitive use of the free circulation of prisoners within prison spaces as an informal means of intimidation in America:

> Nowadays, especially in America, the prison is a reflection of the 'urban jungle'. The public is fascinated by the dangers inherent in prison life … In the past this element was usually instituted from above: forced labor, solitude or the deprivation of liberty. Today, it is being subjected to a dangerous hierarchy among inmates … those in charge of the system use this to scare prospective lawbreakers … In the seventeenth and eighteenth centuries prisoners also mistreated or killed each other, but magistrates would officially present the prison as a salutary environment. Now even policy makers flatly concede what is really going on.

In Spierenburg's account, part of the reason why prison is terrifying and dangerous in America is because inmates are forced to circulate and to socialize with one another, which can expose some to harsh treatment and abuse (see also Milhaud and Moran this volume). The strategic withdrawal of the state's

watchful gaze – as a sort of anti-panoptical strategy – is associated with mobility and personal exposure.

Spierenburg's assertions corroborate an earlier account of the consequences of free prisoner mobility by Engel and Rothman (1984: 428) who associate enhanced within-prison mobility with an increased incidence of violence and rape. They note how one warden who had experimented with increased prisoner mobility, also in an American prison, had recorded that by relinquishing some of his power he 'forced some of the weaker inmates to fend for themselves ... Domination by the benevolent despot warden has been replaced by an environment that permits inmates to hurt each other more than they should'. In both these accounts then, enhanced prisoner circulation within prisons entails increased insecurity and risk for prisoners rather than representing an enhancement of their liberty.

In general, the association of stillness with punishment and mobility with freedom needs to be revisited in the light of the various punitive uses of mobility within prisons. In response to the growing recognition within mobilities studies that mobilities and liberty are not necessarily aligned, the prison studies reviewed here indicate that there are a variety of situations in which prisoner mobility is actually actively enrolled in the curtailment of individual liberties and the delivery of punishments, representing a means by which prisons exert discipline through discomfort, invisibility and exposure to insecurity. These impositions signal the importance of attending to situations in which liberty and mobility are directly at odds.

Electronic Monitoring, Freedom and Punishment

If environments within prisons are increasingly drawing upon mobility in order to be punitive, it is the argument of this section that punishment outside prisons, like electronic monitoring (EM) (but also consider house arrest and community service), often also seek to deliver punishments precisely by maintaining prisoners' physical and social mobility.

Unlike in the case of imposed mobility, however, the increased 'freedom' available under these forms of confinement are, at least initially, attractive to inmates, detainees and their supporters and invite their commitment, their participation and their allegiance. After all, who would refuse to be at home rather than in prison? Yet what transpires is that this apparent freedom functions as a mechanism to reinsert inmates back into productive social relationships, encourage them to reach a point at which they actively desire the responsibilities that prison threatens to remove from them, and more generally to become pliable in the face and promise of a supposed freedom at home.

Seen in this way EM can be seen as a natural extension of mainstream carceral environments, which complements them more than substitutes for them, and pushes the logics of punitive mobility in new directions. To the extent that prison in liberal society seeks to engender a reformed individual, EM operationalizes this

aspiration through a heady mix of new spaces of circulation and new promises of freedom. In this sense EM can be considered as a programme that 'seeks to reconstruct the will of the confined individual in the name of self-control' (Rose 2000: 336). Commentators that account for the rise of EM without noting its governmental potential in these terms (for example by focussing exclusively on the economic case for its proliferation) miss some of its most insipid drivers and effects.

Bauman (2000) argues that the ability to escape from the 'engagements' of social life is the key basis of social power in a globalizing world. In this reading, Bauman understands power as the ability to take "time off" to think and [the] distance allowing a long view' (Bauman 2000: 210). For Bauman, the 'stranger' and the 'vagabond' are the opposite figures to the mobile and elusive 'elite' and can be characterized by an inability to escape and avoid obligations (Bauman 1993). It is in this sense that the maintenance of social mobility through punishments such as electronic monitoring can be seen to burden the convict by forcing them to occupy a position that comes with expectations, duties, responsibilities and mental and physical demands from family, employers, customers, police and even convicts themselves.

These reflections chime with those of prison scholars. As Carnochan (1998: 380) writes, confinement is perfectly possible without imprisonment.

> The prison theme is not encompassed by dungeons, debtors' prisons, penal colonies, internment camps, jails, or penitentiaries alone; such a list, though something of a practical necessity, overlooks the larger, metaphorical pattern that includes all manner of restraint on human action. The overarching category is confinement; its subcategories are captivity of any sort and the particular experience of imprisonment. Confinement restricts the free movement of body and mind.

Carnochan opens the possibility, then, that confinement can be independent of physical restriction, which in turn suggests that forms of punishment that are not explicitly prison-based may be just as constraining, in a qualitatively different sense, as traditional incarceration. For their part, scholars of mobility certainly leave room for the possibility that confinement might not simply refer to physical internment by discussing various different forms of mobility drawing upon Simmel's theorizations of different forms of presence (Simmel and Hughes 1949, Simmel et al. 1997, Urry 2007). Hence, one can be physically present, present in memory, present through one's artefacts and present in the form of a telephone or video-link. This gives rise to the notion that one can be mobile in a corporeal, imaginative, material or virtual way, corresponding to these forms of presence (Urry 2007). It therefore becomes possible to ask how different 'confinement' might look if different combinations of these various mobilities were confiscated. Bearing in mind Carnochan's broad definition of human confinement, the practice of electronic monitoring of suspects, convicts and others might be understood

in terms of the confiscation of different types of presence, marking the end of a fetishization of physical internment in order to achieve confinement.

These insights into the substitutability of different forms of confinement can help us to understand and critique emerging notions of 'equivalence' between different forms of punishment (May and Wood 2005). If corporeal incarceration (i.e. traditional prison) is just one way in which mobility can be confiscated, this raises the question of whether the confiscation of different forms of mobility can achieve 'equivalent' levels of punishment as well as the normative question of whether or not they should be used to do so. For example, could future courts hand down virtual sentences such as barring access to certain internet sites and fora, or destroying virtual identities, not only for virtual crimes but also for crimes in the real world (accepting of course that the distinction between the two is increasingly blurred) (Smyth 2009)? In short, by examining the relationship between different forms of *mobilities* and liberty, responses to the question of the way in which they relate to each other promise to become richer and more nuanced.

EM 'entails the use of radio-frequency or landline telephone technology to know remotely but in real time whether tagged offenders abide by the conditions of a court-imposed curfew' (Nellis 2005: 167). The technology can measure alcohol and drug intake and voice verification monitoring permits the checking of an offender's presence at multiple locations. Satellite tracking using GPS 'monitors movement and facilitates the specification of exclusion zones rather than mere presence at a particular location' (Nellis 2005: 167). More recent innovations also utilize WiFi wireless signals, accelerometers, digital compasses, gyroscopes and altimeters (Yeh 2010).

EM is now widespread. Yeh (2010) estimates that over 225,000 offenders have been electronically monitored in England and Wales alone since its introduction in 1989. This means that over 16,000 people are subject to a variety of forms of EM at any one time in England and Wales (Paterson 2007). The technology has proven popular for a number of reasons. First, it is a cost-effective alternative to traditional prison. The National Audit Office found that keeping an offender on the electronic monitoring system is £70 (US$106) cheaper per day than keeping the offender in prison (Furnham et al. 2010: 1606). In the American context, Yeh (2010) estimates that the large differential between the costs of traditional prison and EM would mean that American society would gain $12.70 (£8.40) for every dollar expended on EM. Second, EM is argued to be effective. Yeh (2010) demonstrates a link between the use of EM and a reduced likelihood of technical violations of court imposed conditions, reoffending and absconding. Its use has also been linked to feelings of enhanced security among those who are otherwise at risk from EM offenders (Erez and Ibarra 2007: 100). Third, EM is also seen as a highly flexible tool of the courts. It 'can be imposed as a condition of bail, as a sentence … or when offenders are released early from prison' (Hucklesby 2009: 250).

Yet EM has been seen as a soft option for criminals. As Killias et al. (2010: 1156) put it, 'many practitioners feel that sitting at home under curfew during nights and leisure-time hours does not require a "real effort" … and may, therefore, not

be sufficiently punitive'. For this reason EM has been associated with a range of 'intermediate sanctions' that locate prison at the harshest extreme of a continuum of sentences available to courts, that also include (in the US), fines, boot camp, house arrest and suspended sentences, many of which allow for a higher degree of convict mobility than traditional prison (Morris and Tonry 1990, May and Wood 2005). The subtext is that punishments that afford volitional corporeal mobility are inherently softer.

An alternative view is put forward by Payne and Gainey (1998, 2004), however, who explicitly reject Morris and Tonry's continuum. They argue that those under EM often experience 'pains of imprisonment' such as the deprivation of liberty and autonomy to a similar degree as those inside prison, and that a range of further impositions can exacerbate the punitiveness of EM in qualitatively different but certainly profoundly acute ways. Through interviews with 27 convicts experiencing electronic monitoring, Payne and Gainey (1998) find that those subjected to EM often experience heightened anxiety and social tension and that they feel acutely watched over and controlled to a similar or greater degree as they might have done in prison (a result corroborated by Paterson, 2007). The 'pains' of EM seem particularly acute when it involves coming into contact with freer individuals frequently, which compounds one's own sense of lack of freedom (that hints at the notion of freedom as a relative, relational concept). The punctuation of everyday life by state involvement that is experienced under EM – such as telephone calls, checking in times and procedures, and the physical notification of the state in the form of a bracelet – is also conducive to a feeling of generalized subjecthood and embarrassment (Payne and Gainey 2004).

Payne and Gainey's arguments concerning the 'pains' associated with EM correlate with other work that demonstrates many convicts' preference to avoid EM when offered the choice, despite the supposed 'freedom' that it offers (May and Wood 2005). In response to a survey question in which 415 prison inmates were asked to indicate whether they thought the statement 'Serving time in prison is less hassle because the [alternatives] have too many responsibilities' was 'not at all important', 'somewhat important', 'pretty important' or 'very important', 26 per cent chose 'very important' and a further 19.8 per cent chose 'pretty important'. Similarly, 22.7 per cent and 13.6 per cent thought that the statement 'In general, living in prison is easier than living outside prison' was 'very important' and 'pretty important' respectively.

We can interpret both the convicts who agreed with these statements and those that did not in terms of the appropriateness of using EM to offer 'free' movement from the perspective of a governmental authority keen to imbue responsible personal habits among soon-to-be liberal citizens. For those that refuse the 'opportunities' of EM, this can be taken to indicate that such criminals are indeed not ready to be reinserted into a liberal order that would rely upon them to embrace their responsibilities. But for those that would choose EM, such convicts do not just represent the over-crowding of the prison system, but the effectiveness of ancillary mechanisms of governmental control that ensure that even the delinquent

individual aspires to be 'better', to improve. This is particularly so among inmates who would chose EM in order to maintain their responsible social roles (employee, father, mother, spouse). For these almost-good liberal subjects, there indeed seems to be a reduced need to incarcerate them at all.

The pains of imprisonment associated with EM – which is a form of punishment that permits a relatively high degree of personal mobility – coupled with the preference for traditional incarceration among many convicts, thus gives us pause for thought in supposing that mobility is desirable or, separately, 'liberating'. Convicts who are more mobile can remain more responsible, since they have not been incapacitated as a result of their punishment. Even accepting the common assumption that mobility entails a degree of freedom, studies of EM show that mobility can function as a way to rule *through* the freedom of subjects (Rose et al. 2006), to responsibilize and ensure that convicts remain answerable to a range of authorities, including family, employers, neighbours and other community officials. While this does not mean that inmates in prison necessarily 'escape' in the way that Bauman intimates, it does problematize the assumed relation between liberty and mobility that is common especially among commentators that view EM as a 'soft option'. Specifically, liberty itself is called into question here as a device that facilitates the smooth running of a liberal governmental order. While the chapter argues that mobility and liberty are often at odds it is clear that 'mobility' here is much less imposed than in the earlier cases discussed in the chapter. Whether this illustrates a weaker governmental imposition or simply a more sophisticated one that draws inmates into the business of carrying out their own punishments and rehabilitation as a function of the freedoms they enjoy is a key issue.

Conclusion: Mobility versus Liberty

The review of penological and sociological studies provided here suggests that scholars of mobility and penology have not gone far enough in prising apart mobility and liberty. Martin and Mitchelson (2009: 460) posit that prisons and detention centres should be viewed as places at which 'human beings [are] 'held' without consent by other human beings'. While there has been recent, and growing, recognition that mobility does not necessarily entail more liberty, the evidence presented here reveals a more overt opposition between mobility and freedom than is frequently acknowledged. This should cause us to expand our understanding of confinement beyond notions of 'holding', or perhaps our notions of holding themselves, to include situations in which people are moved, either against their will or in some cases through their own volition. This chapter has drawn from penological studies, both within prisons and outside them, to establish the ways in which mobility is used to disorientate, hide, terrorize and responsibilize convicts and immigration detainees. Within prisons and prison-like environments, this occurs as a result of the discomfort, invisibility and personal exposure that inmates who are moved around can experience. Outside prisons,

recent evidence concerning the acute 'pains of imprisonment' that non-carceral forms of punishment such as electronic monitoring can achieve point towards the use of mobility as a component form of punishment that responsibilizes, and is perfectly capable of bursting the traditional prison and achieving bitterly acute pains of imprisonment in the process.

The observations of this chapter sit within a wider set of debates about the political uses of mobility. Mountz (2011b), for example, has drawn attention to the mobile state, shifting its borders in order to exclude migrants, or at least render their migration status ambivalent. Amoore (2006) similarly points towards the new responses of states towards increasing human mobility, outlining the regulation of flows of mobility across borders as a key objective of proliferating technologies of security. If political geographers to date have seen the state as reactive in the face of increasingly mobile flows, however, the evidence from penology that has been reviewed here underscores the innovative ways in which states have appropriated mobility in order to actively construct what Rose (2000: 330) has called 'a new archipelago of confinement'.

Of particular note are the ways in which mobilities of inmates and detainees can be seen as governmental techniques (Dean 1999) that are capable of operating both in blunt terms through the disciplining and punishment of delinquents in a governmental order (Foucault 1977), *and*, when they entail freedom, as part of a more sophisticated governmental order that correlates more closely with Foucault's later thinking on governmentality (Foucault 1979). This latter seeks the cooperation of the punished in the delivery of the punishment itself. The prisoner is thus constructed as 'the responsible subject of moral community' who is 'guided – or misguided – by ethical self-steering mechanisms' (Rose 2000: 321).

Overall the chapter has pointed the way towards a fuller understanding of the subjugating and disempowering effects of mobility that work both against and through 'freedom'. This is intended to bolster and further work that rejects the simplistic association between liberty and mobility, or confinement and stasis, and to encourage scholars and practitioners to become more aware of the overtly punitive uses of human movement employed by the modern state.

References

Adams, J. 1999. The social implications of hypermobility. OECD Project on Environmentally Sustainable Transport. Paris: OECD.

Adey, P. 2010. *Mobility*. London: Routledge.

Amoore, L. 2006. Biometric borders: Governing mobilities in the war on terror. *Political Geography* 25(3): 336-351.

Bauman, Z. 1993. *Postmodern Ethics*. London: Routledge.

Bauman, Z. 2000. *Liquid Modernity*. Cambridge: Polity Press.

Bloomfield, B. 2001. In the right place at the right time: Electronic tagging and the problems of social order/disorder. *Sociological Review* 49(2): 174-218.

Bosworth, M. and M.Guild 2008. Governing through migration control: Security and citizenship in Britain. *British Journal of Criminology* 48(6): 703-719.

Braidotti, R. 1994. *Nomadic Subjects: Embodiment and Sexual Difference in Contemporary Feminist Theory*. New York: Columbia University Press.

Burchell, G. 1996. Liberal government and techniques of the self, in *Foucault and Political Reason: Liberalism, Neoliberalism and Rationalities of Government*, edited by A. Barry, et al. London: University College London Press, 19-36.

Carnochan, B.W. 1998. The literature of confinement, in *The Oxford History of the Prison: The Practice of Punishment in Western Society*, edited by N. Morris and D.J. Rothman. Oxford: Oxford University Press, 380-406.

Collins, A. and H. Burns 1992. *Prison Conditions in the United Kingdom*. New York: Human Rights Watch.

Conlon, D. and N. Gill (forthcoming). Asylum seekers and paradoxes of freedom and protest in liberal society. *Citizenship Studies.*

Cresswell, T. 2006. *On the Move: Mobility in the Modern Western World*. London: Routledge.

Cresswell, T. 2010. Towards a politics of mobility. *Environment and Planning D: Society and Space* 28(1): 17-31.

Cruikshank, B. 1999. *The Will to Empower: Democratic Citizens and Other Subjects*. New York: Cornell University Press.

Dean, M. 1999. *Governmentality: Power and Rule in Modern Society*. London: Sage.

Deleuze, G. and F. Guattari 1986. *Nomodology: The War Machine.* New York: Semiotext(e).

Doward, J. 2009. Secrets of CIA 'Ghost Flights' to be Revealed. Sunday 26 July *The Observer*, London, at http://www.guardian.co.uk/world/2009/jul/26/cia-rendition-guantanamo [accessed 13 June 2012].

Ekirch, R.A. 1985. The transportation of Scottish criminals to America during the eighteenth century. *The Journal of British Studies* 24(3): 366-374.

Engel, K. and S. Rothman 1984. The paradox of prison reform: Rehabilitation, prisoners' rights, and violence. *Harvard Journal of Law and Public Policy* 7(2): 431-433.

Erez, E. and P. Ibarra 2007. Making your home a shelter: Electronic monitoring and victim re-entry in domestic violence cases. *British Journal of Criminology* 47(1): 100-120.

Favell, A. 2008. *Eurostars and Eurocities: Free Movement and Mobility in an Integrating Europe.* Oxford: Wiley-Blackwell.

Feldman, A. 1991. *Formations of Violence: The Narrative of the Body and Political Terror in Northern Ireland.* Chicago and London: The University of Chicago Press.

Feldman, A. 2002. X–children and the militarisation of everyday life: Comparative comments on the politics of youth, victimage and violence in transitional societies. *International Journal of Social Welfare* 11(4): 286-299.

Fisher, P. and J. Dobson 2003. Who knows where you are, and who should, in the era of mobile geography? *Geography* 88(4): 331-337.

Foucault, M. 1977. *Discipline and Punish: The Birth of the Prison*. London: Allen Lane.

Foucault, M. 1979. On governmentality. *Ideology and Consciousness* 6: 5-21.

Furnham, A., A. McClelland and E. Drummond-Baxter 2010. The allocation of a scarce correctional resource: Deciding who is eligible for an electronic monitoring program. *Journal of Applied Social Psychology* 40(7): 1605-1617.

Graham, S. and D. Wood 2003. Digitizing surveillance: Categorization, space, inequality *Critical Social Policy* 23(2): 227-248

Hardy Jr, J.D. 1966. The transportation of convicts to colonial Louisiana. *Louisiana History: The Journal of the Louisiana Historical Association* 7(3): 207-220.

Hirschmann, A.O. 1970. *Exit, Voice and Loyalty: Responses to Decline in Firms, Organizations and States.* Cambridge, MA: Harvard University Press.

HM Inspectorate of Prisons. 2006. *Detainees under Escort: A Short Review of Detainees under Escort at Queen's Building and Terminal 3, Heathrow Airport* London: HM Inspectorate of Prisons.

Hucklesby, A. 2009. Understanding offenders' compliance: A case study of electronically monitored curfew orders. *Journal of Law and Society* 36(2): 248-271.

Kaplan, C. 2006. Mobility and war: The cosmic view of US 'Air Power' *Environment and Planning A* 38(2): 395-407.

Killias, M., G. Gilliéron, I. Kissling and P. Villettaz 2010. Community service versus electronic monitoring: What works best? *British Journal of Criminology* 50(6): 1155-1170.

Kofman, E. 2004. Family-Related Migration: Critical Review of European Studies. *Journal of Ethnic and Migration Studies* 30(2), 243-63.

Martin, L.L. and M.L. Mitchelson 2009. Geographies of detention and imprisonment: Interrogating spatial practices of confinement, discipline, law, and state power. *Geography Compass*, 3(1): 459–477.

May, D.C. and P.B. Wood 2005. What influences offenders' willingness to serve alternative sanctions? *The Prison Journal*, 85(2): 145-167.

Medical Foundation for the Care of Victims of Torture. 2004. Harm on removal: excessive force against failed asylum seekers. Granville-Chapman, C., Smith E. and Moloney, N. London: Medical Foundation for the Care of Victims of Torture.

Ministry of Justice. 2008. *Population in Custody: Monthly Tables: November 2007 (England and Wales)* London: Ministry of Justice.

Morris, N. and M. Tonry 1990 *Between Prison and Probation*. New York: Oxford University Press.

Mountz, A. 2011a. The enforcement archipelago: Haunting, sovereignty, and asylum on islands *Political Geography,* 30(3): 18-128.
Mountz, A. 2011b. Specters at the port of entry: Understanding state mobilities through an ontology of exclusion, *Mobilities*, 6(3): 317-334.
Nellis, M. 2005. Electronic monitoring, satellite tracking, and the new punitiveness in England and Wales, in *The New Punitiveness: Trends, Theories, Perspectives*, edited by J. Pratt et al. Cullompton, Devon: Willan Publishing, 167-188.
Nussbaum, M. 2006. *Frontiers of Justice.* Cambridge, MA: The Belknap Press.
Paterson, C. 2007. 'Street-Level Surveillance': Human agency and the electronic monitoring of offenders *Surveillance & Society* 4(4): 314-328.
Payne, B.K. and R.R. Gainey 1998. A qualitative assessment of the pains experienced on electronic monitoring. *International Journal of Offender Therapy and Comparative Criminology* 42(2): 149-163.
Payne, B.K. and R.R. Gainey 2004. The electronic monitoring of offenders released from jail or prison: Safety, control, and comparisons to the incarceration experience. *The Prison Journal* 84(4): 413-435.
Rancière, J. 2004. *The Politics of Aesthetics: The Distribution of the Sensible.* Trans. and introd. Gabriel Rockhill. London and New York: Continuum.
Roberts, J. and T. Gabor 2004. Living in the shadow of prison. *British Journal of Criminology* 44(1): 92-112.
Rose, N. 1999. *Powers of Freedom: Reframing Political Thought.* Cambridge: Cambridge University Press.
Rose, N. 2000. Government and control. *British Journal of Criminology* 40(2): 321-339.
Rose, N., P. O'Malley and M. Valverde 2006. Governmentality. *Annual Review of Law and Social Science*, 2: 83-104.
Schuster, L. 2005. A sledgehammer to crack a nut: Deportation, detention and dispersal in Europe. *Social Policy & Administration* 39(6): 606-621.
Sennett, R. 1998. *The Corrosion of Character: The Personal Consequences of Work in the New Capitalism*. New York: Norton.
Sheller, M. 2008. Mobility, freedom and public space, in *The Ethics of Mobilities: Rethinking Place, Exclusion, Freedom and Environment*, edited by S. Bergmann and T. Sager. Surrey: Ashgate, 25-38.
Shove, E. 2002. Rushing Around: Coordination, Mobility and Inequality. *Lancaster: Department of Sociology, Lancaster University* http://www.lancs.ac.uk/staff/shove/choreography/rushingaround.pdf [accessed 3 December 2010].
Simmel, G., D. Frisby and M. Featherstone 1997. *Simmel on Culture: Selected Writings*. London: Sage.
Simmel, G. and E.C. Hughes 1949. The sociology of sociability. *The American Journal of Sociology* 55(3): 254-261.
Smith, D. 2004. Open borders and free population movement: A challenge to liberalism, in *Spaces of Democracy: Geographical Perspectives on Citizenship,*

Participation and Representation, edited by C. Barnett and M. Low. London: Sage, 113-127.

Smyth, S.M. 2009. Back to the future: Crime and punishment in second life *Rutgers Computer & Technology Law Journal* 36: 18-55.

Sparke, M. 2006 A neoliberal nexus: Economy, security and the biopolitics of citizenship on the border *Political Geography* 25(2): 151-180.

Spierenburg, P. 1996. Four centuries of prison history: Punishment, suffering, the body, and power, in *Institutions of Confinement: Hospitals, Asylums and Prisons in Western Europe and North America, 1500-1950*, edited by N. Finzsch and R. Jütte. Cambridge: Cambridge University Press.

The Economist. 2010. Rough Justice in America, 24 July.

Tonry, M. 2001. Symbol, substance, and severity in Western penal policies. *Punishment & Society* 3(4): 517-536.

Urry, J. 2007. *Mobilities* Cambridge: Policy Press.

Walmsley, R. 2009. *World Prison Population List (Eighth Edition)*. King's College London: International Centre for Prison Studies.

Webber, L. and B. Bowling 2008. Valiant beggars and global vagabonds: Select, eject, immobilize, *Theoretical Criminology* 12(3): 355-375.

Woodward, N. 2006. Transportation convictions during the Great Irish Famine. *Journal of Interdisciplinary History* 37(1): 59-87.

Yeh, S.S. 2010. Cost-Benefit analysis of reducing crime through electronic monitoring of parolees and probationers. *Journal of Criminal Justice* 38(5): 1090-1096.

Chapter 4

Mobility and Power in Detention: The Management of Internal Movement and Governmental Mobility in Romania

Bénédicte Michalon

In response to the important changes migration and asylum policies have undergone in the countries of the European Union, detention facilities for foreigners have been the focus of scholars for the last 20 years. The numbers and types of places dedicated to the confinement of illegal migrants or asylum seekers (detention centres, transit zones at borders and even prisons[1]) have increased both in number and diversity. Lawyers, historians, philosophers and anthropologists examine how these places are connected to the state, its policies and its legislation (Caloz-Tschopp 2004, Darley 2008, Fischer 2007, Julien-Laferrière 1993), and to what extent they stand in continuity or discontinuity with the previous forms of confinement or 'encampment' of foreigners (e.g. Bernardot 2008, Clochard et al. 2004, Intrand and Perrouty 2005, Le Cour Grandmaison et al. 2007). In the early 2000s, scholars began to investigate how space is instrumentalized by states confining foreigners; to what extent detention centres reorganize borders (Clochard 2007, Coleman 2007, Hyndman and Mountz 2008) how their location influences detainees' futures (Mountz 2010) and detention centres' internal spatial organization (Courau 2007).

Only part of this research is based on ethnographic fieldwork carried out inside the places of confinement, mainly because access to detention centres is difficult and very strictly controlled. Most ethnographic writings concentrate on the walls that enclose them; how these walls are constructed, permanently maintained through law and social interactions and how they influence life inside by looking at the way the power of the state is implemented and experienced in detention centres (Darley 2008, Fischer 2007), transit zones (Makaremi 2010), houses of arrest (Le Courant 2010) and also prisons (Boe 2009, Bouagga 2010). Their social, political and administrative functions, which weaken or cut off detainees' relationships (Le Courant 2010) and information networks with the outside world (Fischer 2005) are highlighted. However, others argue that the walls of detention centres are just as porous as prison walls (Darley 2008), following Grosser (1960) and Ohlin (1960),

1 The types of places of confinement for foreigners vary from one country to another, which makes it difficult to establish a complete list.

who considered prisons not just as closed spaces but in relation to the outside (Chantraine 2004, Combessie 1996, Renouard 1999, Wacquant 2002).

My own fieldwork in Romania convinced me of the necessity to further analyse the relationships between confinement and mobility.[2] There are two detention centres for foreigners awaiting deportation in Romania. A country known for the emigration of its own citizens, has itself become a country of transit and immigration over the last 20 years (Michalon and Nedelcu 2010). The opening of its borders in 1989, its 2007 accession to the European Union and the ongoing process of accession to the Schengen Area have pushed it to the forefront of the European and global migration scene.

Reinforcement of the management and control of international migrations coincide with negotiations for the European integration of Romania in 1997. The first detention centre, located in Otopeni, a northern suburb of Bucharest close to the country's main airport, was opened in 1999 and renovated and extended in 2005. The second centre, in the small village of Horia close to the city of Arad and the Romanian-Hungarian border, opened in 2003 (see Figure 4.1). Before 1999 Romania had no laws regarding detention: prior to the establishment of the centres, foreigners arrested at the border or without documents were confined inside Otopeni Airport (CPT 1998, 2003).

Detention in one of these 'centres for public custody' (*centre de custodie publică*) is 'a temporary measure of movement restriction on the territory of the Romanian State'[3] taken by a magistrate against foreigners awaiting deportation from Romanian territory. Both are 'closed places, designed and laid out for temporary accommodation',[4] run by the Romanian Office for Immigration (*Oficiul Român pentru Imigrări*, ORI), part of the Ministry of the Interior. Foreigners detained in these centres can be divided into three categories: irregular migrants without documents to enter or remain in the country, foreign former prisoners convicted of a crime, and 'undesirable' foreigners considered a 'threat' to national security and public order.[5] State control over foreigners therefore relies on both confinement and 'imposed' mobility (removal). The maximum duration of detention is six months (or two years for former prisoners) and can be extended to up to 18 months under certain circumstances.

---

2 This research is part of the TerrFerme programme *Mechanisms of confinement. A territorial approach to contemporary political and social control* (financed by the French National Office for Scientific Research and by the Aquitaine Regional Council: http://terrferme.hypotheses.org/). The author thanks the Romanian NGO she worked with and the TerrFerme team for the collective work which enabled her to develop the ideas presented here.

3 Urgency ordinance of the Romanian Government no.194/2002 on the foreigners' regime in Romania; Law no.157/2011 to modify and complement certain normative acts on the foreigners' regime in Romania.

4 Ibid.

5 Romanian law distinguishes between removal (*indepărtare*) from the Romanian territory, which is an administrative measure, and deportation (*expulzare*), which is a penal one.

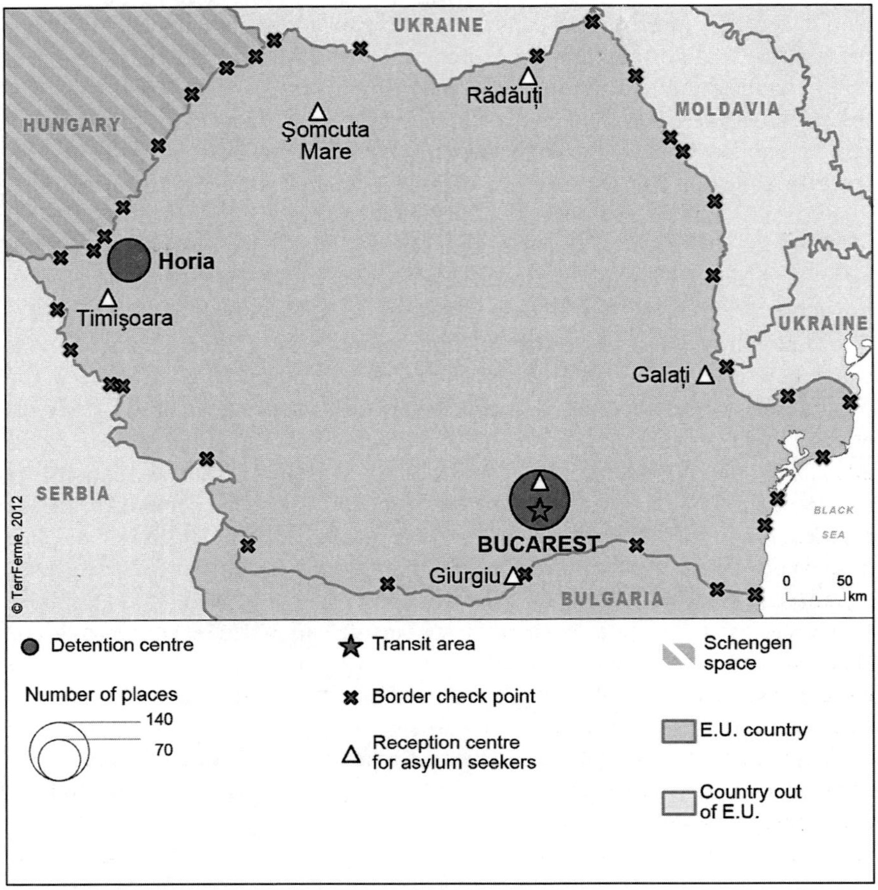

Figure 4.1 Detention estate for foreigners in Romania

Fieldwork revealed the importance of the movements between the different places that make up the 'detention universe' (detention centres, police offices, offices of institutions in charge of migration, consulates etc.), both in Romania and at a European level. The interviews conducted with detainees and former detainees clearly showed the lasting effect of the tension such movements create. These interviews as well as observations within the detention centres also highlighted the stakeholders involved in circulation inside the centres, and in its control.

This paper draws upon recent work which develops the idea that movement is an integral part of confinement (see Gill 2009a, 2009b, Pallot 2005, Philo 2001[6])

6 See also the sessions 'Geographical Political Economies of Incarceration', *Annual International Conference, Royal Geographical Society*, London, September 2010; and

and, at a broader level, defines confinement in terms of its relation to mobility: 'we restrict our definitions of "imprisonment" and "detention" to intentional practices that (i) restrict individuals' ability to move from one place to another and (ii) impose orders of space and time so that individual mobility is highly constrained, if not eliminated' (Martin and Mitchelson 2009: 460); 'detention and prisons may serve as a "spatial fix" for surplus capital and labour ... but this fixing relies on global transportation networks and global financial flows' (ibid.: 472).

Moran et al. (2012) stress that much research which characterized the 'mobility turn' (see e.g. Adey 2006, Cresswell 2010, Cresswell and Merriman 2011, Sheller and Urry 2006, Urry 2007) only considers voluntary mobility, overlooking forced mobility; characterized by the limited autonomy of those who are moved. Combining research on mobility and confinement makes it possible to bring into focus aspects of mobility that have remained barely analysed until now. Up to what point is mobility a technique used for the government of populations? Using the Romanian case, the chapter argues that mobility is central to power relationships in detention. Power is understood here using a Foucauldian approach as the product of a multiplicity of spatialized social mechanisms intended to discipline individuals, without them necessarily realizing it (Foucault 1975) – and not a system of institutions. The analysis takes place at two levels: at the level of internal mobility, managed in a very rational way in the detention centres themselves, and at the level of external mobility as an integral part of confinement. The chapter thus contributes to Gill's (2009a) concept of 'governmental mobility'; 'a governmental technique that acts upon this ostensibly powerful population [asylum seekers in the UK]' (2009a: 187), by identifying two types of governmental mobility: the first bound to punitive measures against some of the detainees, the second related to detention and removal.

Methodology

Data were generated during ethnographic fieldwork carried out in Romania in 2009 and 2010, using observation and semi-structured interviews. Observation was realized *in situ*, within the two Romanian detention centres. Access was time-limited (one to two days for each visit), and I was interviewed by the directors at the beginning and end of my visits. I was strictly forbidden to use any sound recording or photography – the detainees themselves are not allowed to have a camera or a mobile phone with camera. Once in the centre, I remained under surveillance and could not move freely. I could visit the different parts of the centres – administrative offices, checkpoint of the wardens, collective spaces for the detainees, refectory – but was always 'accompanied' by a policeman. I could enter 'witness' cells, but had to remain on the threshold of inhabited cells,

'Mobilities, Borders and Confinement: Confinement as Mobility', *Association of American Geographers Annual Meeting*, Seattle, April 2011.

in order to 'respect the privacy' of their occupants. In one of the detention zones, I could circulate alone in the central corridor, but with a checkpoint located in this corridor, I remained permanently under the wardens' gaze.

I was allowed to interview a dozen foreigners within the detention centres. My respondents were not selected by the staff, but some of the interviewees apparently kept good relationships with the policemen. It was not possible for me to select the interviewees in order to build a representative sample. The first obstacle was language; I carried out the interviews in Romanian, English, French or German. I tried to talk to foreigners with various experiences (duration of stay in Romania, duration of stay in detention, former prison or confinement experience). In one centre, I carried out interviews in a staff meeting room; in the other centre, I interviewed detainees in the visiting room; requested to leave the door open, 'for my safety'. Security cameras video recorded conversations.

The majority of fieldwork was undertaken outside the detention estate during six months volunteering in an NGO assisting detainees and former detainees. I was allowed to interview the 'beneficiaries' of the NGO; about 30 foreigners released from detention, usually talking with them several times, alone, in the NGO premises. Some agreed to be audio-recorded; the majority refused. I also carried out about 15 interviews with asylum-seekers previously detained in Romania or elsewhere in Europe.

The Management of Circulation in Detention Centres as an Aspect of, and Evidence for, Power Relationships

At first glance, a detention centre is a space of non-circulation: detainees spend most of their time confined to their cells. They are not allowed to leave the centre, except for their legal proceedings or for hospital care. The first level of mobility is therefore the cell; 12 square metres, designed to accommodate four people, where movements are restricted to walking from the bed (where detainees spend most of their time) to the bathroom (part of the cell) and the window, used to communicate from one cell to the next and to exchange small objects. However, movements inside the centres do exist. The question is where, when and how they take place. While the architecture of the facilities, the time schedules and the statuses of the detainees are used as tools by the institutions to manage the internal flows of movement, the detainees do preserve some 'room for manoeuvre'.

A Divisive and Mobility-Restricting Architecture

The architecture of Romanian detention centres is characterized by numerous barriers and boundaries (a feature also typical of prisons – see Milhaud 2009, Dirsuweit 1999), aiming first and foremost to control and to restrict internal flows.

The surrounding walls of the centres are the first obstacle to mobility. In Bucharest, the high walls have barbed wire, video surveillance and police security

checks on arrival and departure. In Horia the lower surrounding wall has no wire, the gate is not locked and identities are only checked on entry.

This first barrier encloses many more, which create different and separate detention areas and zones. While each unit has facilities for general services such as administration (director's office, secretariat, control room etc.), laundry, catering and medical care (infirmary, psychologist's office), these are strictly separated from the zones of confinement, with detainees occupying a limited part of the buildings. This division impacts on circulation within the centres; it is anything but easy to pass from the 'world of staff' to the 'world of detainees'.

The detainees' zone in Bucharest is evocative of typical communal accommodation. The reinforced front door gives access to a narrow central corridor with the cells on one side and the communal rooms – visiting room, laundry room, the office for an NGO, a so-called 'sports room' and 'library' – on the other. A space in the middle of the corridor (next to the security surveillance screens) has armchairs and a television for the wardens, who spend most of their time here. The corridor leads to a large fenced terrace, connected via a footbridge to the dining room in the other building. The centre has space for 164 people, held in cells with barred windows, reinforced doors with spyholes, and alarm bells connecting each cell to the checkpoint. Two punishment cells are situated at the end of the corridor.

The centre in Horia, which can accommodate up to 60 detainees, is organized in panoptical style, with cells located *vis-à-vis* the guards' office, from which they are separated by two grid doors. The gangway leading to the cells is enclosed and the only communal room on the ground floor is equipped with a television, a table-tennis table and a coffee machine. The cells resemble those in Bucharest.

Both centres share typical features of carceral architecture: cells but no dormitories, lack of privacy and circulation spaces, numerous barriers and control devices (reinforced doors, spyholes, locks, barbed wire and fences, video surveillance systems, alarms …). The centres differ, however, in the location of the wardens as the representatives of power. In Bucharest, they work in the middle of the corridor; detainees cannot avoid seeing them, being seen, or meeting them when allowed to leave their cells. By contrast, the Horia guards work outside the detainees' area, and contact is only visual.

In summary, the architecture of the detention centres is an efficient tool of control, creating spatial divisions and specific zones for each group and organizing the flow of movements. Movement, however, is further limited by both time schedules and the status of the detainees.

The Control of Circulation by means of Time and Status

The management of circulation by time creates a regular daily routine, just like in a prison (Kantrowitz 1996). The daily schedule is clearly a tool to control the mobility of the detained. The rules in Bucharest restrict detainees' freedom of movement inside their zone. Apart from during mealtimes, they are allowed to leave their cells for only three hours per day. Whilst in theory, they should be

allowed into the courtyard of the centre every day, in reality this depends on the presence and willingness of wardens to surveil them, and on whether detainees are deprived of their time outside as a punishment (as in the summer of 2009 after the escape of a Georgian detainee). The time schedule in Horia is more flexible; detainees are allowed to leave their cells and to move around the gangway every day from 7am to 8pm. After lunch they spend one hour outside in the courtyard and after that up to three hours in the common room. Three times per day the detainees eat in the canteens, located in other parts of the building. Other than that they leave their zone only in exceptional circumstances and if escorted by wardens (for example to talk to the director or to the medical staff, in the administration area). The barriers between the detainees' area and the staff zones can only be crossed with permission, reasserting the asymmetrical power relationship between detainees and staff.

The right to leave the cell also depends on the status of the detainees, who are divided into categories and distributed accordingly into cells and buildings. This procedure reflects both official and unofficial categories, e.g. irregular migrants, former prisoners, men, women or families etc. on the one hand, and sociocultural groups, social networks, age, relationships with the wardens etc. on the other. In both centres men are separated from women and families,[7] while irregular migrants are supposed to be separated from former prisoners. Whenever possible, detainees perceived by the staff to share cultural and/or religious similarities share cells.

The imperative is to prevent men from meeting women and families, whether in the detainees' areas or in the courtyard. Ideally, there should be no interaction between these groups in the canteen, the administrative zone or the courtyard. Hence, women and families are placed in specific zones: in a separate building in Bucharest and on the lower floor in Horia. The management also prevents contact between irregular migrants and former prisoners. In theory, the schedules of both groups in Bucharest accommodate this; one group leaves their cells in the morning, the other in the afternoon.[8]

Internal movements are generally restricted to the detainees' zone, but may also extend into the staff area. While the detainees are allowed to move freely around their zone while cells are open, any access to the administrative area is strictly controlled by the police. A detainee wanting to talk to the management or to another member of staff such as a doctor or a psychologist needs to catch the warden's attention and ask for his request to be transmitted. This is not always easy. As the alarm bells in the cells in Bucharest do not always attract the officers' attention, detainees frequently bang on their doors to be heard. In Horia the grid doors at the ends of the gangway are places of contact between the detainees and the wardens; if a detainee has a request, a group will gather at the doors. At mealtimes, the stairs leading to the canteen are opened, wardens channel the movement of

---

7 This spatial division exists mainly in theory, as only very few women and families are detained in Romania. Most immigrants to the country are men.

8 The institution rules in Horia do not seem to take account of this parameter.

the detainees, count them, and oversee the canteen during the meal. A detainee entering the administrative area is permanently under surveillance. The 'border' between the administrative area and the detention zone itself is a reinforced door, which can only be opened by the wardens on watch inside the detainees' area. If someone wants to enter – be it the director of the centre or a visitor – they need to ring the bell and wait for a police officer to open the door.

Tactics and Room for Manoeuvre

These rules, however, can be subject to tactics (De Certeau 1998) and arrangements both on the part of the police and the detainees. This is especially the case for the so-called 'intermediaries' who act as a link between the detainees and the police officers, entertain privileged relations with the staff and therefore occupy a prominent position among the detainees. They communicate information, intervene in negotiations between both parties and transmit individual or collective requests from the detainees to the staff and vice versa. There are 'rewards' for this informal role; in Bucharest the two intermediaries were allowed to leave their cell and move freely around the detainees' area from breakfast until dinner, avoiding one of the aspects of detention described as most distressing by the detainees: restriction to the cell for most of the day. Their informal role also earned the intermediaries the right to undertake domestic tasks inside the centre, such as cleaning communal areas like the terrace and corridor. Taking rubbish to the dustbins in the courtyard was considered a real privilege, offering detainees the opportunity to spend a few extra minutes outside and to pick fruit from a tree in the grounds of a neighbouring farmhouse.

Although the organization of mobility in the centres is first and foremost the responsibility of the management, the detainees influence the structure of their movements, such as their use of the communal areas like corridors and 'leisure' rooms. In Bucharest, they mainly use the corridor to stretch their legs, filing in silence from one end to the other and back again. They spend little time in the common rooms – the 'library', 'sports room' and terrace – which are sparsely equipped and uninviting, and the courtyards, which are bleak and austere. The areas dedicated to common use are poorly equipped for this purpose, leading to the conclusion that the strict control over internal mobility in the detention centres goes hand-in-hand with lack of effort to make the common rooms attractive to the detainees. If visits to other cells are not prohibited by internal regulations, then they are subject to unofficial rules. Wardens consider the cells private areas and do not enter unless necessary or unless invited in by the occupants. Similarly, detainees will only enter another cell if invited to do so. These official and unofficial rules of circulation reveal much about the relational dynamics and internal social hierarchies prevalent amongst the detainees, and to which the police contribute to a certain extent.

Circulation and its management inside Romanian detention centres is a major issue for the institutions. Since the same observation has been made in detention

centres in France (Fischer 2007), one can assume that the management of internal mobility constitutes a major technique of control over detainees, just like in a prison (Milhaud 2009). It is hence possible to argue that detention – like imprisonment (Kantrowitz 1996, Martin and Mitchelson 2009, Philo 2001) – is characterized by paradoxical relations between confinement and controlled – if not coerced – mobility. This first level of connection between confinement, mobility and power inside, is complemented by a second, outside the centres.

Moving Detainees for the Purposes of Confinement: Governmental Mobility

Detention can at first glance appear to be a temporary means of immobilizing people. Its ultimate objective, however, is to remove foreigners. (Coerced) mobility is therefore an integral part of the concept of detention. The Romanian Office for Immigration operates this type of circulation and carries out its task under conditions that reassert the asymmetrical power relationship between foreigners and authorities. 'Governmental mobility' is the procedure to which foreigners are submitted, but it can also serve as a punitive measure and ultimately results in detainees being deprived of their migration route.

Mobility as Part of the Procedure

A foreigner in detention is not immobile and confined in one single place. A detention order is followed by a series of movements which are organized by the authorities. The movement of detainees for procedural purposes takes place at both national and European levels. In Romania it is strictly controlled by the Romanian Office for Immigration (ORI). It is common practice to take detainees to the consulate(s) which is/are likely to grant the required pass. If a foreigner appeals against the decision of detention (which is generally the case), his case is heard in court. Even if these movements are an integral part of the detention routine, they cause organizational difficulties. Due to the lack of escorting staff, wardens working in the detention centres are sometimes deployed for these tasks – to the detriment of the remaining detainees. Things become particularly complicated if a foreigner detained in Horia needs to be escorted to Bucharest, usually resulting in a transfer to the detention centre in Bucharest for a few days or weeks. The conditions of these transfers are extremely trying for the foreigners: firstly, they generally receive no information about when and why they are to be moved. They are transported in a police van escorted by several policemen, often handcuffed and sometimes without any food or drink during the 12-hour journey. On departure and arrival they are submitted to a full body search. Appearing in court handcuffed is experienced as an act of unjustified violence by people who do not consider themselves criminals. The entire organization of these transfers can therefore be understood as a way for the authorities to reassert the power relationship.

Transfers for procedural purposes also take place at a European level. European migration control tools enable the forced transfer of foreigners between EU Member States and to partner states outside Europe. Two mechanisms are of particular importance; the first is the readmission agreements signed between EU Member States and extra-European countries. By 2010 Romania had signed 29 so-called 'bilateral agreements', and was a signatory to all the agreements between the EU and a third country ('community agreements'). These are the legal basis for foreigners without documents to be moved from one country to another – regardless of which country they are citizens. The second mechanism is the so-called Dublin II Regulation, which came into effect in September 2003. Its purpose is to prevent asylum seekers from filing an application for asylum in more than one Member State, and to determine which Member State is responsible for examining the application where an asylum seeker has passed through several EU countries. Usually, the State responsible is the state through which the asylum seeker first entered the EU. The implementation of the readmission agreements and the Dublin II Regulation relies on *Eurodac*, a European system for comparing the fingerprints of asylum seekers and illegal immigrants in the EU, which helps identify the countries through which a foreigner has passed or in which he has filed for asylum in order to transfer the migrant concerned to the state which agrees to admit him under either of the two systems. The aim of the readmission agreements and the Dublin II Regulation is the coerced displacement of foreigners for the purpose of a 'distribution of tasks' between Member States. However, the removal of immigrants beyond or in the absence of readmission agreements does exist, which shows how important the coerced displacement of foreigners is in today's 'management' of migration.

Removals can only be carried out successfully if migrants remain static and accessible by the relevant authorities. Their confinement in detention centres, which prevents them from 'escaping' on the one hand and helps to implement their removal on the other, is the solution applied in most European countries, although other solutions such as electronic tags exist elsewhere. Confinement itself implies mobility, since foreigners cannot always be detained close to their place of residence or arrest. In Romania, both the detention centres and the transit zones at the land and air borders[9] enable the surveillance of foreigners awaiting expulsion.

Removal itself takes place under very coercive conditions. Foreigners without documents or asylum seekers who have passed through Romania and have been arrested in another European country usually try to avoid returning to Romania. They use all legal means in order not to be removed and frequently attack their own bodies through hunger strikes or self-mutilation. The removal procedure itself passes through a series of detention facilities in several countries with highly variable conditions, and upon their arrival in Romania migrants are

9 The maximum duration of confinement in the transit zones for foreigners applying for asylum at the Romanian borders is 20 days.

re-confined (with the exception – in some cases – of asylum seekers). This 'travelling detention' or 'mobile detention' provides migrants with a detailed knowledge of confinement conditions in different countries. The way the police officers treat them and the material conditions in detention are used in 'comparative analysis', allowing detainees to re-evaluate their situation on arrival in Romania.

Punitive Mobility

Transfers of detainees from one Romanian detention centre to the other can also take place for punitive purposes, independent of deportation procedure, for example if the behaviour of a detainee is considered problematic by wardens, if someone appears to be vindictive, for example, or to be a leader of protest movements inside the centres, or if someone is suspected of carrying out illegal activities or of maintaining relations with human traffickers outside. These punitive transfers seem to be the result of rather arbitrary decisions, since the place of detention is in theory chosen by the prosecutor. Decisions about transfers, however, are made by the ORI outside the official procedures and are generally a trying experience for the detainees concerned.

The punitive transfer of a detainee to another centre is not only a simple act of transportation. Its objective is to influence the person transferred, as has been shown in respect to the transportation of prisoners (Svensson and Svensson 2006); to change the detainee's behaviour to bring it in line with the expectations of the police. The transfer is also an opportunity for the institution to reassert its power: the social structure within a centre is disrupted by the departure or arrival of a (new) detainee and the relationships between detainees, and even between detainees and wardens, are disrupted. The detainee concerned suddenly finds himself in a new social environment to which he needs to adjust, and which isolates and weakens him (see Gill 2009a, 2009b). The transfer itself is yet another occasion for the authorities to reassert their power, the conditions under which it takes place being just as challenging as during a transfer for procedural purposes.

Punitive mobility is clearly an integral part of control over detained foreigners and very similar to the transfer of prisoners (see for example Milhaud 2009). Gilles Chantraine (2004: 125) considers it an 'institutionalized impossibility': punitive transfers organized by prison administration aim 'at breaking any attempt to rebuild relationships and prepare a plan for a new attempt [of escape], which leads to loneliness, suffering and specific deprivations'.

Given the impact of both procedural and punitive mobility on detainees and their psychological state of mind, one may ask whether governmental mobility as imposed on detained foreigners is not a punishment *in itself*, just like the transportation of Russian prisoners to their colonies (Piacentini et al. 2009).

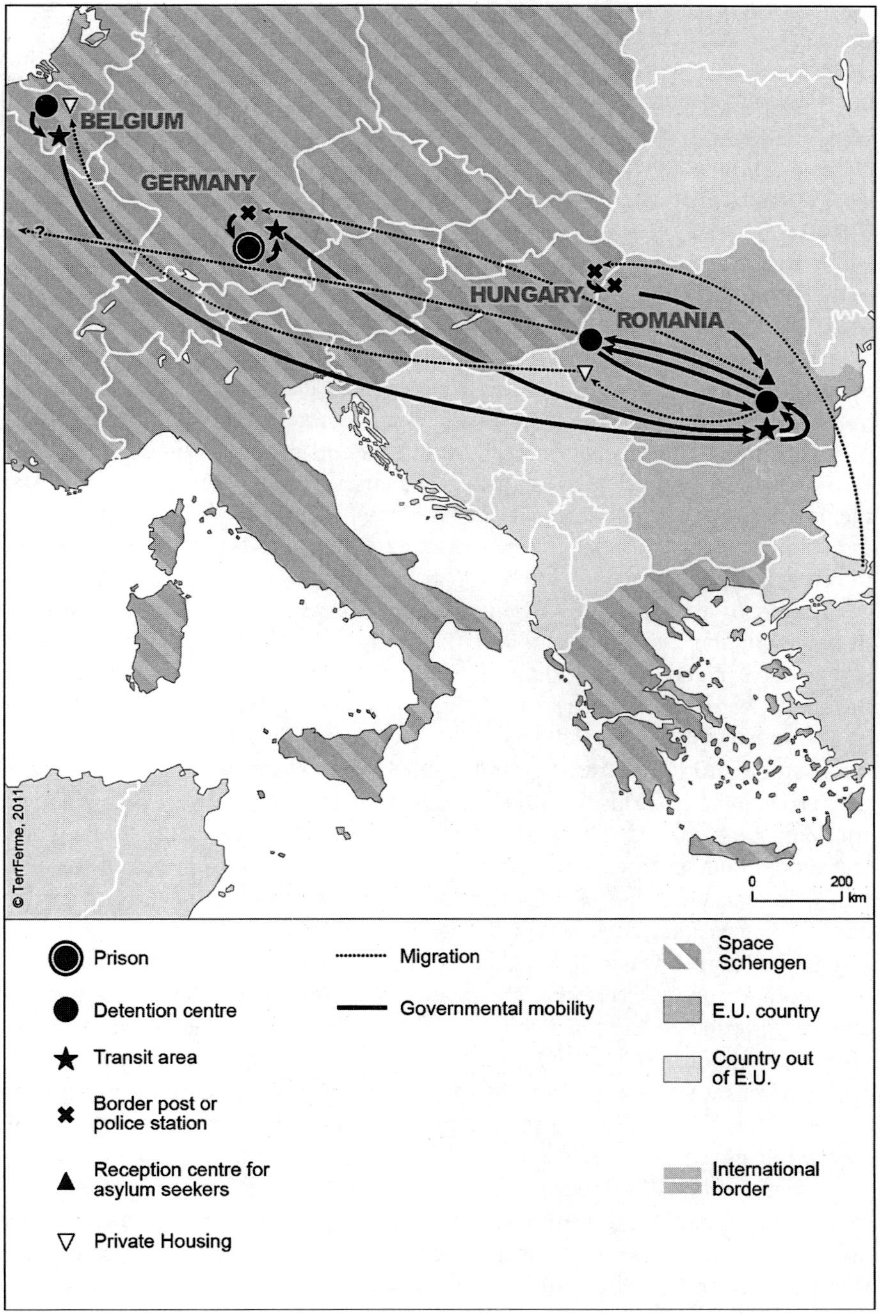

Figure 4.2 Migration route of an Asian migrant and former detainee in Romania (2008–2010)

Self-deprivation and Deprivation of the Migration Route

Some researchers argue that the experience of detention changes the way foreigners look at themselves and the way they act in their immigration country. Detention makes them aware of their status as so-called 'illegal immigrants'. In this sense, they are deprived of themselves and their self-representation is reduced to illegality (Le Courant 2010). This assumption can also be analysed from a spatial point of view: detention not only deprives immigrants of their sense of themselves, but also of their migration route. This is illustrated by the story of a Sri Lankan immigrant who passed through detention in Romania (see Figure 4.2).

He arrived in the country in 2008 and was arrested during an attempt to cross the border to Hungary. After spending one night at a border check-point, he was taken back to Romania. Having spent two days in a closed facility of the Romanian border police, he applied for asylum in Romania and was placed in a centre for asylum seekers, where he lived for eight months. While his application for asylum was examined, he left the country for Germany, but was arrested by border police and spent 24 hours at a police station. He was then taken to prison in Munich, where he stayed for two months for having used false documents. Whilst being taken to the airport for deportation back to Romania, he self-harmed by cutting his veins, but his injuries did not prevent him from being flown out the following day. On arrival in Romania, he spent 24 hours in the transit zone of the airport without being able to leave the cell in which he was confined. He was then taken to the detention centre in Bucharest, where he stayed for 10 days, before being transferred to Horia and subsequently being taken back to Bucharest for his hearing at the court of appeal. He never received any explanation as to why he had been taken to the other end of the country under extremely difficult conditions: handcuffed, without his mobile phone and without any food or drink. After his release, he spent five months in a private room in the city of Timişoara and then left for Belgium, where he lived and worked for five months before being arrested once again. He spent 25 days in a detention centre and undertook a hunger strike in an attempt to escape another deportation to Romania. The Belgian police eventually managed to take him to the airport, where he spent one day in a detention cell before disembarking. On arrival in Bucharest, he spent another day in the transit zone of the airport before transfer to the detention centre. After two weeks he was taken back to Horia, where he stayed for five and a half months. After his release he filed another application for asylum.

By spring 2010, he had spent most of the two previous years in various detention facilities for migrants and asylum seekers in Hungary, Romania, Germany and Belgium. His migration route was determined by European regulations (the readmission agreements and the Dublin II Regulation) and the forced transfers organized by the Romanian Office for Immigration. As shown in Figure 4.2, his route has a circular shape and sheds light not only on the geography of mobility, but also on the geography of detention and, on a broader scale, of confinement.

Detention is supposed to end with the deportation of the foreigner concerned or, if the latter fails, with his release. Either procedure is another occasion for the authorities to reassert their power. Being released does not necessarily mean that detention is over; according to Romanian regulations, an immigrant can only be placed in detention once per motive of arrest (illegal entry on the Romanian territory, use of false identity documents etc.), but it is not prohibited to put someone in detention several times. Such 'round-trip tickets' between detention and liberty are quite common for migrants without documents who do not manage to obtain a residence permit. Hence, the 'circularity' of the migration routes on a European level is complemented by the circular movements within Romania of migrants unable either to obtain documents or to leave the country.

Getting accurate information on how expulsions are carried out is difficult. The statements as regards their preparation as well as their execution are very contradictory. The ORI is responsible for release from the centres, for transfer to the airport and for removal. According to staff, releases are prepared in so far as the detainees are informed about the day and the time of their deportation. This information is even said to be used as a means of pacifying detainees. From a police point of view, a 'good' detainee is a detainee who accepts the decision of his expulsion. According to detainees and former detainees however, the police officers use expulsion as a means for psychological pressure; threats of expulsion are used in order to control the behaviour of detainees and removal itself occurs by surprise: there is no information about the day or time of departure. Hence, even the end of detention is characterized by the power relationship between the authorities and the foreigners.

Release can occur quickly, if an appeal against the detention of a migrant (or against its prolongment) is successful. Information about the date of the release is part of the tools used by the police officers in order 'to ease the tensions of everyday life', as the director of one of the centres put it. However, little is done to prepare a detainee for release. On the 'big' day, he receives his personal belongings, the gate is opened and he finds himself alone without anyone to help him find a place to go or somewhere to stay. Occasionally, the centre management will contact an NGO and ask for temporary accommodation, but such efforts largely depend on the 'vulnerability' of the detainee and his relationship with staff during detention. Removal and its failure consequently result in the confinement of foreigners in a space of forced mobility (Makaremi 2008: 60), in a 'ping-pong' between their country of origin, their countries of transit and their country of 'arrival', in which they are anything but welcome.

Conclusion

The spatial dimension of confinement is not only a matter of fences. The question is what role mobility plays in the system of administrative detention of migrants. The Romanian example highlights how important the circulation of the detainees

within the facilities is for the institution. Their mobility is strictly controlled through architectural features, the time schedule and the distribution of different categories of detainees within the buildings. A little room for manoeuvre, however, does exist and manifests itself in informal hierarchies among detainees, and unofficial relationships with the wardens. In detention centres, just as in prisons, the rational management of mobility is therefore an instrument of control. Analysis of the Romanian case also demonstrates that mobility arguably plays a more important role in detention centres than in prisons, due to the transfers of detainees, organized by the Romanian and European authorities. These transfers are an integral part of the procedures, which, on a broad scale, include the mechanisms of the 'distribution' of migrants within the EU and certain partner states, as well as the mechanism of removal from Romanian territory. This procedural mobility on both national and international levels also determines the power relationships between the authorities and the foreigners. This power is even more evident when the purpose of these governmental transfers is punitive. Eventually, the detainees are deprived of their migration route and ultimately of their capacity to decide on and control their own life. One can conclude that the various types of mobility within the detention system play a determining role in the asymmetry between the police and foreigners. Movement is used to govern and control people and, in this sense, represents an ultimate form of deprivation of space and hence of self-deprivation (Le Courant 2010).

Concerning the analysis of forced mobility, detention in Romania shows the diversity of techniques of government through mobility: in their causes, forms, scales, spaces and places. The biography and migration paths of detainees and former detainees show that 'free' mobility and forced mobility follow one another throughout the detainees' trajectories, as they live a succession of movements that they decide on, and others that they undergo without consent.

Considering forced mobility and detention also has consequences for the way detention is perceived. Until now, spatial analyses focussed how places of confinement were controlled and supervised, on what happened inside, (particularly as regards the appropriation of space), and on how these facilities were integrated into the social and economic environment and used for territorial policies. Therefore, relating mobility to detention is an argument for moving from a static vision of confinement towards a conceptualization of it 'in motion', with much fuzzier boundaries than prison walls (Baer and Ravneberg, 2008, Comfort 2003, Moran 2011). Confinement is also located in the spaces of mobility. Thinking of mobility as an integral part of, and as caused by, confinement shows that it is necessary to leave the centres in order to change the focus and understand the system in its entirety. The works of Olivier Razac on electronic tagging offer a stimulating analysis (Razac 2009, 2010a, 2010b, and Gill this volume). This tag, used by the French prison authority as an alternative to imprisonment, serves to monitor the movements of the convicted and, according to Razac, radically changes the spatial dimension of confinement. He argues that the tag, which is often considered a virtualization of confinement, 'is an addition rather than a

substitute for traditional prison facilities' and that it 'establishes new limitations in open space' and 'intensifies existing spatial constraints' (Razac 2010a: 215). In this sense, it 'revives and extends the carceral management of space' (ibid.: 209). One can argue that reflections on the control of closed spaces may evolve towards an analysis of the control of spatial practices.

References

Adey, P. 2006. If mobility is everything then it is nothing: Towards a relational politics of (im)mobilities. *Mobilities* 1: 75-94.

Baer, L.D. and B. Ravneberg 2008. The outside and inside in Norwegian and English prisons. *Geografiska Annaler* 90(2): 205-216.

Bernardot, M. 2008. *Camps d'*étrangers. Bellecombe-en-Bauges: Editions du Croquant.

Boe, C. 2009. Parcours de détenu, parcours de migrant. La prison productrice d'irrégularités, in *Enfermés dehors*, edited by C. Kobelinsky and C. Makaremi. Bellecombe-en-Bauges: Éditions du Croquant, 209-226.

Bouagga, Y. 2010. Rentrer dans le droit commun? *Champ pénal / Penal field, nouvelle revue internationale de criminologie* [Online], VII. Available at http://champpenal.revues.org/7898 [accessed 30 May 2012].

Caloz-Tschopp, M.-C. 2004. *Les* étrangers *aux frontières de l'Europe: le spectre des camps.* Paris: La Dispute.

Chantraine, G. 2004. *Par-delà les murs.* Paris: PUF.

Clochard, O., Y. Gastaut and R. Schor 2004. Les camps d'étrangers depuis 1938: continuité et adaptations. Du modèle français à la construction de l'espace Schengen. *Revue Européenne des Migrations Internationales* 20(2): 57-87.

Clochard, O. 2007. Le jeu des frontières dans l'accès au statut de réfugié en France: une géographie des politiques européennes d'asile et d'immigration. University of Poitiers: PhD in Geography.

Coleman, M. 2007. Immigration geopolitics beyond the Mexico-US Border. *Antipode* 39(1): 54-76.

Combessie, P. 1996. *Prisons des villes et des campagnes.* Étude *d'*écologie *sociale.* Paris: Éditions de l'Atelier.

Comfort, M.L. 2003. In the tube at San Quentin. The 'Secondary Prisonization' of women visiting inmates. *Journal of Contemporary Ethnography* 32(1): 77-107.

Courau, H. 2007. *Ethnologie de la forme-camp de Sangatte. De l'exception à la régulation.* Paris: Éd. des Archives contemporaines.

Cresswell, T. 2010. Towards a politics of mobility. *Environment and Planning D: Society and Space* 28(1): 17-31.

Cresswell, T. and P. Merriman 2011. *Geographies of Mobilities: Practices, Spaces, Subjects.* Farnham: Ashgate.

CPT (European Committee for the Prevention of Torture and Inhuman or Degrading Treatment or Punishment). 1998. *Report to the Romanian Government on the*

CPT's visit to Romania from 24 September to 6 October, 1995. Strasbourg: Council of Europe.

CPT (European Committee for the Prevention of Torture and Inhuman or Degrading Treatment or Punishment). 2003. *Report to the Romanian Government on the CPT's visit to Romania from 24 January to 5 February 1999.* Strasbourg: Council of Europe.

Darley, M. 2008. Frontière, asile et détention des *étrangers*. Le contrôle *étatique* de l'immigration et son contournement en Autriche et en République tchèque. Institut d'Etudes Politiques de Paris: PhD in Political Sciences.

De Certeau, M. 1998. *L'invention du quotidien.* Paris: Gallimard.

Dirsuweit, T. 1999. Carceral spaces in South Africa: a case study of institutional power, sexuality and transgression in a women's prison. *Geoforum* 30(1): 71-83.

Fischer, N. 2005. Clandestins au secret. Contrôle et circulation de l'information dans les centres de rétention administrative français. *Cultures et Conflits* 57: 91-118.

Fischer, N. 2007. La rétention administrative dans l'Etat de droit. Genèse et pratique du contrôle de l'enfermement des étrangers en instance d'éloignement du territoire dans la France contemporaine. Institut d'Etudes Politiques de Paris: PhD in Political Sciences.

Foucault, M. 1975. *Surveiller et Punir. Naissance de la prison.* Paris: Gallimard.

Gill, N. 2009a. Governmental mobility: The power effects of the movement of detained asylum seekers around Britain's detention estate. *Political Geography* 28: 186-196.

Gill, N. 2009b. Longing for stillness: The forced movement of asylum seekers. *M/C Journal*: A *Journal of Media and Culture* 12(1). Available at http://www.journal.media-culture.org.au/index.php/mcjournal/issue/view/still [accessed: 30 May 2012].

Grosser, G.H. 1960. External setting and internal relations of the prison, in *Theoretical Studies in Social Organisation of the Prison*, edited by R. Cloward et al. New-York: Social Science Research Council, Pamphlet 15, 130-144.

Hyndman, J. and A. Mountz 2008. Another Brick in the Wall? Neo-refoulement and the externalisation of asylum by Australia and Europe. *Government and Opposition* 43(2): 259-269.

Intrand, C. and P.A. Perrouty 2005. La diversité des camps d'étrangers en Europe: présentation de la carte des camps de Migreurop. *Cultures & Conflicts* 57: 71-90.

Julien-Laferriere, F. (ed.) 1993. *Frontières du droit, frontières des droits: l'introuvable statut de la 'zone internationale'*. Paris: L'Harmattan.

Kantrowitz, N. 1996. *Close Control. Managing a Maximum Security Prison. The story of Ragen's Stateville Penitentiary.* New York: Harrow and Heston.

Le Cour Grandmaison, O., Lhuilier, G. and Valluy, J. (ed.) 2007. *Le retour des camps? Sangatte, Lampedusa, Guantanamo....* Paris: Autrement.

Le Courant, S. 2010. Ce que fait la politique de contrôle de l'immigration. De l'étranger menotté au clandestin. *Champ Pénal / Penal Field, Nouvelle revue*

internationale de criminologie, VII. Available at http://champpenal.revues.org/7889 [accessed: 30 May 2012].

Makaremi, C. 2008. Pénalisation de la circulation et reconfigurations de la frontière: le maintien en zone d'attente. *Cultures et Conflits* 71: 55-73

Makaremi, C. 2010. Zone d'attente pour personne en instance. Une ethnographie de la détention frontalière en France. University of Montreal: PhD in social and ethnological Anthropology.

Martin, L.L. and M.L. Mitchelson 2009. Geographies of detention and imprisonment: Interrogating spatial practices of confinement, discipline, law, and state power. *Geography Compass* 3(1): 459-477.

Michalon, B. and M. Nedelcu 2010. Histoire, constantes et transformations récentes des dynamiques migratoires en Roumanie. *Revue d'Etudes Comparatives Est-Ouest* 41(4): 5-27.

Milhaud, O. 2009. *Séparer et punir. Les prisons françaises : mise à distance et punition par l'espace*. Université Michel de Montaigne Bordeaux 3: PhD in Geography.

Moran, D. 2011. Between outside and inside? Prison visiting rooms as liminal carceral spaces. *GeoJournal*, Online First, 3 December 2011. Available at http://www.springerlink.com/content/p805836842373543/ [accessed 30 May 2012].

Moran, D., L. Piacentini and J. Pallot 2012. Disciplined mobility and carceral geography: prisoner transport in Russia. *Transactions of the Institute of British Geographers* 37: 446-460

Mountz, A. 2010. *Seeking Asylum. Human Smuggling and Bureaucracy at the Border.* Minneapolis: University of Minnesota Press.

Ohlin, L.E. 1960. Conflicting interests in correctional objectives, in *Theoretical Studies in Social Organisation of the Prison*, edited by R. Cloward et al. New-York: Social Science Research Council, Pamphlet 15,111-129.

Pallot, J. 2005. Russia's penal peripheries: Space, place and penalty in Soviet and post-Soviet Russia. *Transactions of the Institute of British Geographers* 30: 98-112.

Philo, C. 2001. Accumulating populations: bodies, institutions and space. *International Journal of Population Geography* 7(6): 473-490.

Piacentini, L., J. Pallot and D. Moran 2009. Welcome to Malaya Rodina ('Little Homeland'): Gender and penal order in a Russian penal colony. *Social and Legal Studies* 18(4): 523-542.

Razac, O. 2009. Le bracelet électronique et la virtualisation de l'enfermement. *Cultures et Sociétés* 10: 52-57.

Razac, O. 2010a. Le placement sous surveillance électronique mobile (PSEM): un nouvel espace de la peine ? *Cahiers sur la sécurité* 12: 209-215.

Razac, O. 2010b. *Le placement sous surveillance* électronique *mobile: un nouveau modèle pénal?* Report of the Ecole National d'Administration Pénitentiaire: Agen.

Renouard, J.M. 1999. *La Prison de l'Île-de-Ré: un travail d'équipe.* Guyancourt: Questions pénales, Bulletin du CESDIP, XII (4).

Sheller, M. and J. Urry 2006. The new mobilities paradigm. *Environment and Planning A* 38(2): 207–226.

Urry, J. 2007. *Mobilities.* Oxford: Polity Press.

Wacquant, L. 2002. From slavery to mass incarceration: Rethinking the 'Race Question' in the US, *New Left Review* 13: 41-60.

Laws

Urgency ordinance of the Romanian Government no. 194/2002 on the foreigners' regime in Romania (Ordonanţa de urgenţă a Guvernului nr.194/2002 privind regimul străinilor în România), in UNHCR. 2008. Legislaţia naţională relevantă în domeniul azilului şi migraţiei. Bucharest: Editura C.H. Beck.

Law no. 122/2006 on asylum in Romania (Legea nr.122/2006 privind azilul în România), in UNHCR. 2008. Legislaţia naţională relevantă în domeniul azilului şi migraţiei. Bucharest: Editura C.H. Beck.

Law no. 157/2011 to modify and complement certain normative acts on the foreigners' regime in Romania (Legea nr.157/2011pentru modificarea şi completarea unor acte normative privind regimul străinilor în România).

Chapter 5
'You don't even know where you are': Chaotic Geographies of US Migrant Detention and Deportation

Nancy Hiemstra

Introduction

Numerous scholars studying immigration enforcement have pointed to the significance of particular spatialities of immigration enforcement. For example, Bloch and Schuster (2005) identify detention, deportation and dispersal as largely symbolic tactics employed in the United Kingdom to control migrant bodies. Martin and Mitchelson (2009: 469) note that the 'uneven topography' of migrant detention is tied to 'uneven geographies of legal practices and boundaries'. Mountz (2004: 336, 2010) discusses the Canadian government's 'long tunnel thesis', showing how 'detached geographies of detention' are strategically used to limit refugee claimants' access to resources that could support their cases. Collyer (2007) and Hyndman and Mountz (2008) detail emerging strategies of 'externalization' countries are implementing to prevent migrants, including asylum seekers, from ever entering their territory. Gill (2009) explores ways in which frequent transfers of detained migrants in the UK negatively shape the perceptions of influential actors in detention centres. Similarly, Martin (2012) has shown how, to the US Department of Homeland Security (DHS), difficulty of access is a positive trait when selecting detention sites.

In this chapter I augment this scholarship by visually and descriptively mapping elements of detention and deportation that are central to the experiences of migrants detained in the United States. To do so, I draw on fieldwork conducted in Ecuador with the families of migrants detained in the US and migrants deported from the US I elaborate in particular on the time and space geographies of detention paths as detainees undergo frequent yet unpredictable transfers. Through attention to the uneven, illogical time-space geographies often experienced by migrants, I suggest that the detention and deportation system (hereafter D & D system) produces myriad *chaotic geographies*.

Chaos is defined as 'a state of utter confusion' and *chaotic* as 'completely confused or disordered' ('Chaos' 2001). A 'chaotic geography' of the D & D system, I propose, is a particular way in which chaos is spatialized and temporalized.[1]

1 Mountz (2010: xxix) has previously used the term *chaotic geographies* in a different, if related, meaning: 'Borders exist as walls only in the geographical imagination. In the

I posit that through attention to these chaotic geographies one can scrutinize the driving forces behind the enactment of immigration policy in the United States. Importantly, the chaos ensures the continued profit and influence of the immigration industrial complex, a term used to describe the 'confluence of public and private sector interests in the criminalization of undocumented migration, immigration law enforcement, and the promotion of 'anti-illegal' rhetoric' (Golash-Boza 2009b: 296; cf. Fernandes 2007). As Eyal Weizman (2007: 8) writes regarding the intentionally disordered settlement strategy pursued by the Israeli government, 'Chaos has its peculiar structural advantages'. Through my mappings of detainees' experiences, I scrutinize the ways in which the immigration industrial complex both capitalizes on and contributes to the chaotic geographies of detention and deportation.

The chapter is organized as follows. First, I provide brief background information for Ecuadorian migration to the US and my research methodology. Then, I explore the influence of the immigration industrial complex in shaping the rapidly growing detention system, including material conditions of detention and the development of internal economies in detention facilities designed to profit on detained migrants. Next, I literally and figuratively map detainees' spatial and temporal experiences of detention by focusing on the ICE's practice of frequently transferring detained migrants. Through these various mappings, it becomes clear that the chaos inherent to the US D & D system functions to the advantage of the immigration industrial complex.

Ecuadorian Migration to the United States

There is a long history of migration from Ecuador to the US, beginning in the 1950s and greatly increasing in the late 1990s with a severe national economic and political crisis (Acosta et al. 2004, Gerlach 2003).[2] Current scholarly calculations usually place the number of Ecuadorians in the United States around 500,000 to 600,000 (Gratton 2007, Jokisch and Kyle 2008) – a significant percentage of Ecuador's total population of around 14 million. Approximately 70 per cent of Ecuadorians in the United States live in the New York City area. Ecuadorians have also settled in the cities of Chicago, Columbus (Ohio), Los Angeles, Miami, Minneapolis, and suburban areas around Boston, New York and Philadelphia (Kyle and Jokisch 2005, FLACSO 2008, Jokisch and Kyle 2008). As in other nations experiencing high levels of migration, these patterns of human mobility have profoundly impacted the Ecuadorian economy, politics, culture, communities and families.

The number of Ecuadorians deported from the United States has grown in accordance with the US government's increasing emphasis on this immigration

---

practice of border enforcement, the border is enacted through more dispersed, chaotic geographies.'

2 These crises also spurred significant migration from Ecuador to Europe, primarily Spain and Italy (Jokish and Kyle 2006).

enforcement tactic. According to Department of Homeland Security (DHS) data, 2,321 Ecuadorians were deported from the US in 2010, ranking it as number seven on ICE's list of countries to which immigrants were deported (DHS Office of Immigration Statistics 2011). Though small in comparison to the total number of deportees for 2010 (387,242), the number of Ecuadorian deportees is noteworthy when Ecuador's population size (14 million) is compared to that of similarly-ranked countries: number six is Brazil, with 3,190 deportees and a population of almost 200 million, and number eight is Colombia, with 2,267 deportees and a population of 45 million.

The bulk of fieldwork in Ecuador was conducted over nine months in 2008 and 2009 in the southern Andean city of Cuenca, an epicentre of Ecuadorian migration to the United States. Primary methods were participant observation and semi-structured interviews. I volunteered four days per week at Cuenca's municipally-run House of the Migrant. There, I helped Ecuadorian family members locate and obtain information regarding migrants detained in the US. In doing this work I used the computer program Skype to call dozens of ICE offices and detention facilities, private detention centres, and county and city jails. In addition, I interviewed 40 Ecuadorians after their deportation from the US regarding their experiences of detention and deportation, as well as their post-deportation realities and plans. Interviewees were contacted through family members who had sought assistance. Four interviewees were women and the rest men.[3] Interviewee ages ranged from 20 to 47, with an average of 30. Participants had been deported between April 2008 and August 2009. Twenty-two of the interviewees had been caught in the US interior, while 18 were apprehended at or near the US-Mexico border.

The Immigration Industrial Complex and the D & D System

The immigration industrial complex, which consists of entities that somehow benefit from increasing immigration enforcement and anti-immigrant sentiment, has played a critical role in the post-9/11 emphasis on detention and deportation in US policy. Throughout the 1980s and 1990s, rates of criminal incarceration in the US grew rapidly, spurred by the growing influence of private prison corporations and businesses that supply prisons in various ways, or what has been called the 'prison industrial complex' (Davis 2003). The increasing emphasis on detention in immigration policymaking occurred in tandem with this expanding prison system, as growing numbers of local jails and private corporations also embraced detention as a source of revenue (Welch 2002). Also within this complex are politicians and media personalities who gain personal advantage by emphasizing and reinforcing immigrants' illegality (Welch 2002, Fernandes 2007, Golash-Boza 2009b).

In this maelstrom of anti-immigrant sentiment and racialized fears, immigrants and refugees 'become raw materials for the corrections business' (Welch 2002: 166). As Hernández (2008: 47) succinctly puts it, 'detention has become a

3 Roughly 20 per cent of all deportees from the US to Ecuador are women.

growth industry for private corrections companies and a development strategy for local municipalities seeking to contract jail space with the federal government'. Indeed, immigrant detention can be a profit machine (Welch 2002). The two largest immigrant detention companies in the United States, the Geo Group and the Corrections Corporation of America, earn annual profits of $1.17 billion and $1.69 billion, respectively (Cervantes-Gautschi 2010). Many counties have also come to see immigrant detention as a desirable source of profit, working with private companies to construct new facilities (Barry 2009a 2010). In addition, there are a wide range of other kinds of companies involved, from phone companies earning profits on immigrants' calls, to service companies providing food, to physicians providing medical services (Fernandes 2007, Golash-Boza 2009a, Barry 2010). Furthermore, there are millions of individuals employed in the immigrant detention and deportation industry who have a personal stake in its perpetuation (Welch 2002).

Lobbying by private companies with a stake in detention has played an instrumental role in directing the development of DHS policies, including policies requiring increased detention capacity (Fernandes 2007, Barry 2009b, Cervantes-Gautschi 2010; Sullivan 2010). For example, lobbying played a pivotal role in DHS's 2006 policy shift regarding its pursuit and prosecution of undocumented immigrants. Instead of being charged with the misdemeanour misuse of a social security number, arrested immigrants were suddenly charged with crimes such as identity theft and document forgery (Cervantes-Gautschi 2010). For immigrants (both documented and undocumented), such crimes – as well as any others that have possible sentences of a year or more – fall into the category of 'aggravated felonies'. Aggravated felony is a distinct category of crime for which immigrants (not US citizens) face mandatory detention and deportation, with few opportunities for relief even in circumstances such as extended residence in the US or being parent to US citizen children. This policy shift massively expanded the potential detention population and, consequently, opportunities for profiting from detention (Cervantes-Gautschi 2010).

In addition, lobbying efforts contribute to the maintenance of weak regulatory frameworks (Flynn and Cannon 2009), allowing detention providers to increase profit by providing a minimum of services for detainees. The average daily cost to the government for detaining a migrant is $122 (Detention Watch Network 2010). Largely due to successful lobbying, there are few standards in place regarding conditions of migrant detention (Fernandes 2007, Flynn and Cannon 2009). Numerous reports have shown that detainees typically experience poor conditions including inadequate food, substandard medical care and unsanitary facilities. Interviewees also described tight control over facility-issued items such as clothing, toiletries and bedding, findings that correspond to recent investigative reports (Barry 2009a, American Friends Service Committee (AFSC) 2010, Bernstein 2010a, 2010b, ACLU 2011).

The commodification of migrant bodies extends beyond the 'per head' payments for housing detainees. The D & D system has nurtured the development of economies internal to detention facilities (Wood 2011). For example, detained

migrants' outside communication is largely controlled by private phone contractors who charge exorbitant rates. While outside facilities one can call between Ecuador and the US for as low as nine cents per minute, deportees reported rates ranging from $5 to $30 for five minutes. In addition, detention centres usually sell food, recreation-related items such as cards and toiletries to detainees at inflated prices. At the same time, detainees labour within facilities for absurdly low wages. Some detention facilities cut costs by 'employing' detainees for tasks such as cleaning, food preparation and laundry (Wood 2011). According to DHS guidelines, no immigrant detainee can receive more than $1 per day for labour performed (Fernandes 2007). Therefore, facilities pay little for labour that otherwise would earn at least minimum wage. Despite the demeaning wage, many of my interviewees were grateful, even desperate, for the opportunity to earn small amounts of money in order to purchase calling cards or extra food. Several interviewees also mentioned bartering with other detainees in the near-absence of wages, such as teaching English in exchange for extra food. Thus, an internal economy is crafted that operates cyclically. The poor conditions of detention create demand for certain purchases, which makes migrants willing to work for next to nothing, and then they return the little that they earn back to the facility economy via their purchases.

I argue that the immigration industrial complex is tightly linked to the chaos of the US D & D system, as described below. While the chaos may be partly attributed to the rapid growth in detention and the securitization of immigration law since the 1990s (particularly since 9/11/2001) (Kanstroom 2007a, 2007b, Coleman 2007a, 2007b, 2009), additional scrutiny reveals that the chaos is not entirely as careless as it may seem. That is, the lack of cohesion and transparency within the D & D system is highly effective in obscuring migrants' alternatives to deportation. In addition, due to its disorder and confusion, the D & D system projects a cloak of impenetrability that hides both its faults and the driving influence of the immigration industrial complex. I explore these relationships in the following empirical accounting of detainees' spatial and temporal experiences.

Space and Time Geographies of Detention and Deportation

In this section, I focus on the centrality of space and time in the chaotic structures of detention and deportation. Before elaborating my empirical work on detainee transfers, I briefly review several key factors shaping detainees' experiences, in addition to the basic material conditions of detention mentioned in the previous section.

Temporal Uncertainty and 'Undue Processes'

First, an increasing volume of work documents the lengthy, unpredictable detention durations detainees often experience, particularly if they are attempting to fight deportation (Welch 2002, Dow 2004, AFSC 2010, ACLU 2011). While Immigration and Customs Enforcement (ICE) states that migrants spend an

average of 30 days in detention, investigative reports indicate this number is actually significantly higher. Due to mandatory detention laws in combination with severe court backlogs, those migrants who are waiting to see a judge (often to fight deportation orders) face extended detention (ACLU 2011, AFSC 2010).[4] Waiting for necessary travel documents can also delay deportation. Once ICE has the paperwork, the migrant must wait for a deportation flight, either on JPATS (the Justice Prisoner and Alien Transportation System, operated by the US Marshals) or commercial flights. Second, detainees are often subjected to 'undue processes' (Hernández 2008: 48) that work to rigidly maintain them on the path to expulsion. The 'undue processes' that are common to detainees' experiences include a lack of legal representation, lack of translation services, and signatures consenting to deportation obtained by threat and bullying (Miller 2003a, Human Rights Watch (HRW) 2009, Coutin 2010, National Immigrant Justice Center (NIJC) 2010, ACLU 2011). Third, detainees' access to information and communication is rigidly controlled. Reports document how detainees may not be informed of rights they do have or avenues for fighting deportation (AFSC 2010, NIJC 2010, ACLU 2011). Clearly, detained migrants' experiences are profoundly shaped by extended timelines made more frustrating by curtailed rights, isolation and uncertainty. Now, I focus on the chaotic geographies of detainees' transfer paths.

Mapping Detention Paths

Migrants detained in the US D & D system experience a high degree of forced mobility (Welch 2002, Fernandes 2007, Hernández 2008, HRW 2009, TRAC 2009, ACLU 2011). A recent report notes that as of 2008, the numbers of transfers that take place per year exceed the number of detainees. Over 50 percent of detainees are transferred at least one time, and one in four detainees is transferred multiple times (TRAC 2009).

In my interviews with deportees, I asked them to name in order the places where they had been detained, and for how long they were detained. Frequent mobility was experienced by many of the 40 interviewees, for whom the average number of transfers was 3.4. Ten were detained in more than four places. One interviewee was transferred eight times. Detention durations ranged from two and a half to 28 weeks. From this data, I constructed six maps to visually represent detainees' transfer paths, which provide a starting point for further discussion of the chaotic time-space geographies of migrant detention.[5]

4 As of June 2011, the average wait time for Immigration Court was 482 days (TRAC 2011). While this backlog is due partly to ICE's goal of raising deportation numbers (Preston 2011), it also stems from a failure to keep the courts adequately staffed (TRAC 2010, 2011).

5 Cartography is by Joseph Stoll, Syracuse University Department of Geography.

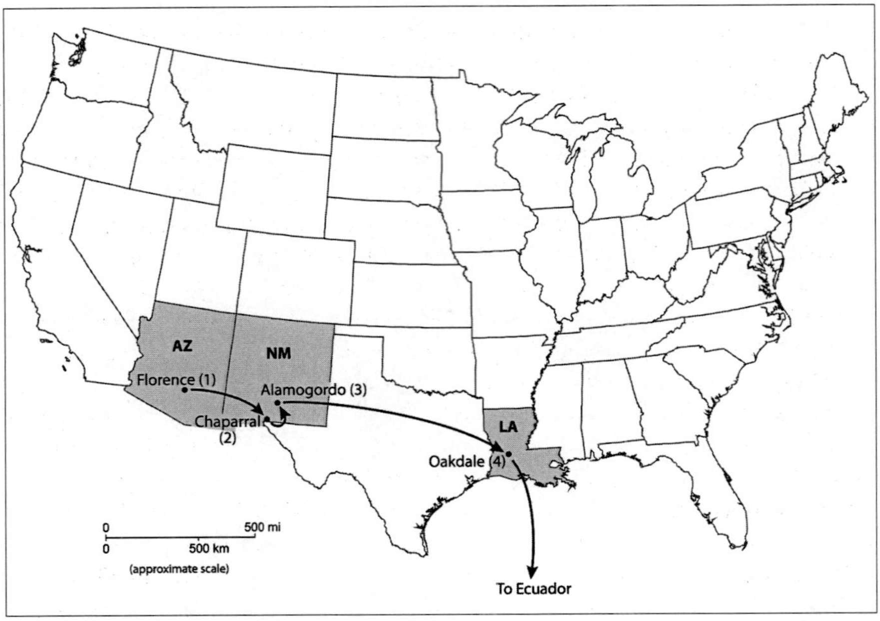

Figure 5.1 Detention transfer path – Eduardo

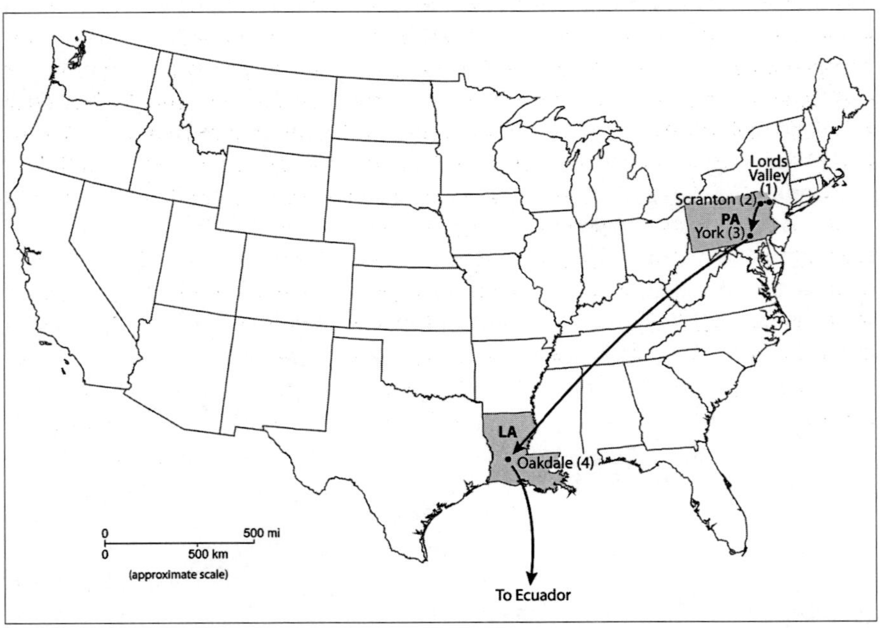

Figure 5.2 Detention transfer path – Remigio

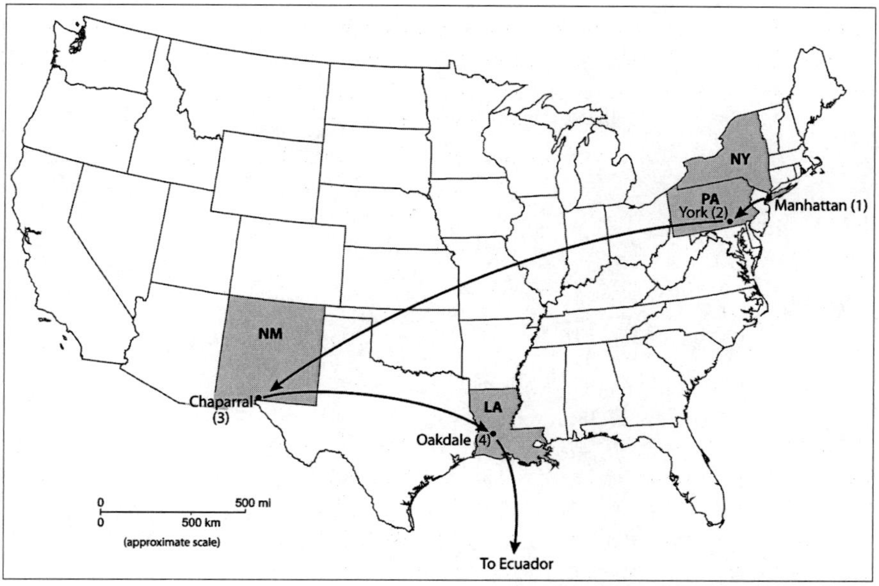

Figure 5.3 Detention transfer path – Hugo

The first three maps illustrate the paths of migrants transferred four times. Figure 5.1 shows the transfer path of Eduardo. After his capture near the US-Mexico border in Arizona, Eduardo was detained in four facilities over the course of eight weeks. Eduardo's transfers were relatively straightforward, keeping him generally in border states and moving him closer to the Louisiana facility from which many interviewees' JPATS flights departed. The transfer history of Remigio is also fairly uncomplicated (Figure 5.2). In a period of six weeks, Remigio passed through three facilities in Pennsylvania, and then he was moved to Louisiana for a JPATS flight. While Hugo was also detained in four places, the map of his transfers shows a more geographically inefficient route (Figure 5.3). During his six weeks in detention Hugo was moved inland from Manhattan to Pennsylvania, then flown west two-thirds of the way across the country to New Mexico, then back east to Louisiana for deportation.

The second three maps depict the paths of migrants detained in more than four places, which was the experience of one quarter of my interviewees. Marcelo's path, shown in Figure 5.4, illustrates the geographical absurdity inherent to many transfers. Detained in a total of six places over 12 weeks, Marcelo zigzagged and looped through space before being deported. He was first detained in New Jersey, and transferred to Manhattan and then to York, Pennsylvania. Next, he was moved to Raymondville, Texas, then to Laredo, Texas, and then to Louisiana from where he was deported. Some patterns of

forced mobility were so complicated as to make mapping them a challenge. Oscar, for example (Figure 5.5), was transferred seven times over 10 weeks, and one of the transfers moved him back to a facility where he had been previously. He was first detained in Queens, New York, then moved to Manhattan, and then to Freehold, New Jersey. His fourth transfer brought him back to Manhattan before being moved to York, Pennsylvania, then to El Paso, Texas, and finally to Laredo. Diego, like numerous other interviewees, did not know the names or locations of several facilities through which he passed (Figure 5.6).[6] Diego was detained for 12 weeks and transferred eight times. He was first held in Manhattan, then moved to a facility somewhere in New Jersey, then another place he could not name in Pennsylvania, and then to York, Pennsylvania. Next, he was transferred to San Antonio, Texas, then Raymondville, then to a facility in New Mexico of which he did not know the name or location. Diego was then transferred to Louisiana and from there deported. I now draw on these maps as well as data from other interviews with deportees to further examine the forced mobility within the D & D system.

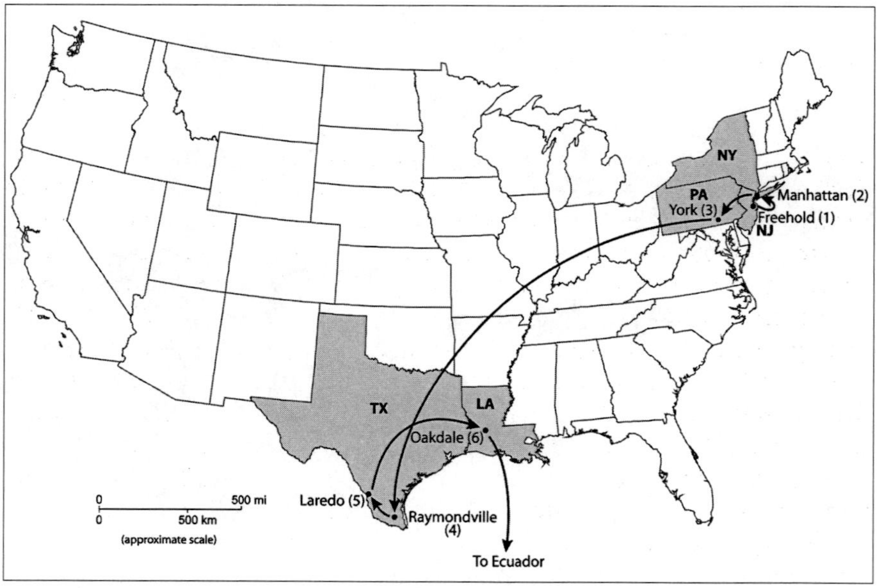

Figure 5.4 Detention transfer path – Marcelo

6 For the detention locations about which this interviewee was uncertain, I mapped points representing known sites of detention in the general area indicated by the interviewee.

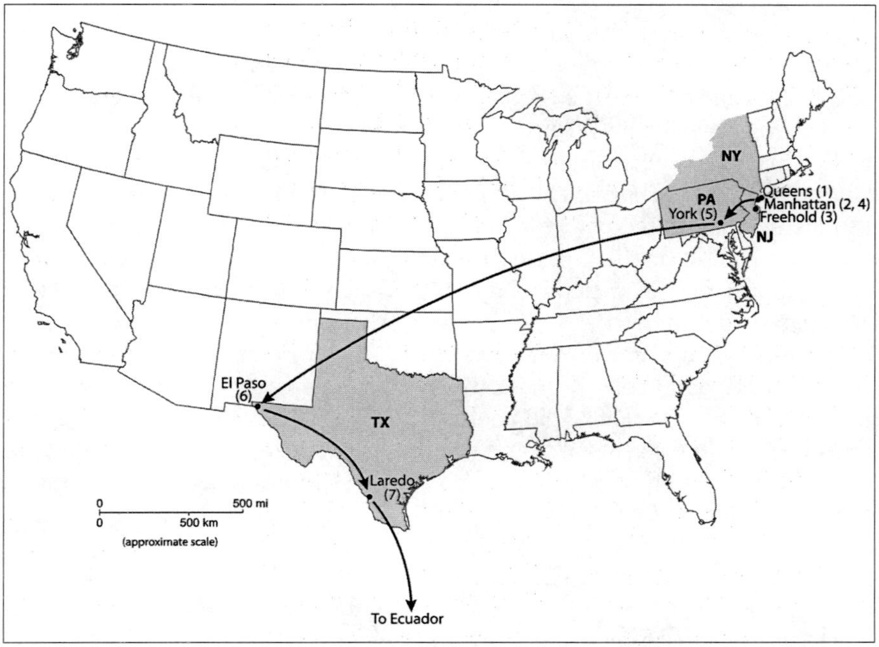

Figure 5.5 Detention transfer path – Oscar

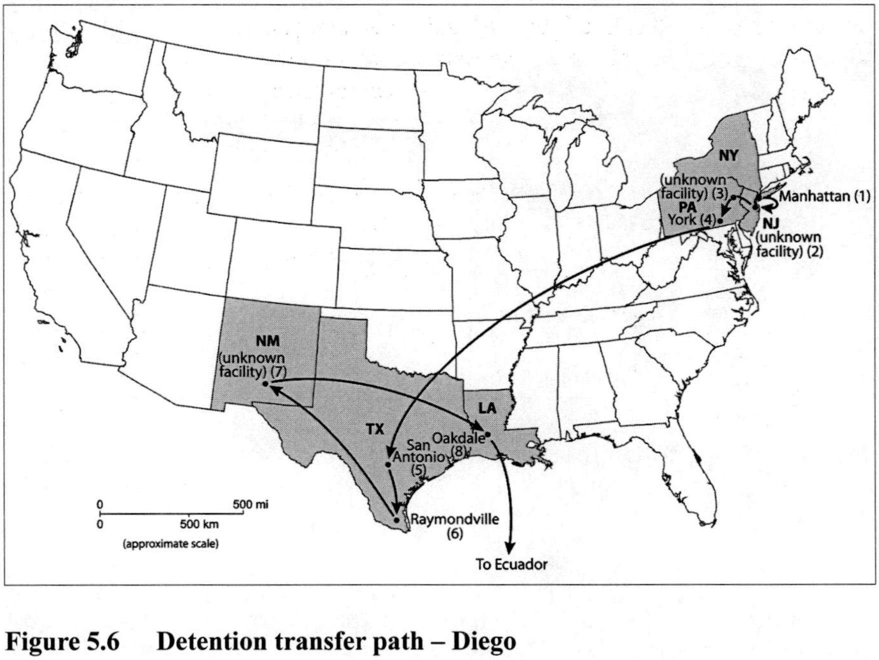

Figure 5.6 Detention transfer path – Diego

Deciphering Transfer Patterns

Some of these transfer geographies follow a comprehensible calculus. For instance, five of the six final transfers illustrated in the maps were to Louisiana, from where deportation flights often depart. Some facilities, such as those in Manhattan and York, Pennsylvania, are hubs through which most migrants in a particular region generally pass. Interestingly, where migrants are initially apprehended appears to loosely correspond with the number of places in which they are then detained. Migrants caught near the border usually pass through the lowest number of facilities, while the greatest number roughly correlates with initial detention in the interior.[7] This disparity makes some sense when one considers typical apprehension to deportation trajectories. Migrants caught on the border trying to enter the US are less likely to enter complicated procedural trajectories (such as fighting removal orders based on claims of family ties or length of residence), and they are already close to common JPATS departure points. In contrast, those caught in the US interior, like Marcelo, Oscar and Diego, are far from the US-Mexico border and are more likely to have complicated, time-consuming procedural trajectories. Likewise, the shortest detention periods were most likely to be similar to that of Julio César, detained for a total of two and a half weeks in two places. He was apprehended near the US-Mexico border, had no previous entry attempts on record, and received the necessary paperwork in time for the first available deportation flight to Ecuador. The longest detention periods were usually those of migrants fighting a removal decision. This was the case for Saira, who was detained for 16 weeks, and for Sergio, detained for 28 weeks.

Still, the time and space geographies of migrants' detention paths often do not adhere to discernible formulas. For example, despite the fact that Eduardo's (Figure 5.1) situation was similar to that of Julio César's, he was detained for significantly longer (eight weeks) and experienced two more transfers. Movements did not necessarily bring a migrant closer to a departure point for flights, such as Diego's (Figure 5.6) transfer from Raymondville to New Mexico before being transferred to and deported from Louisiana. Martín was transferred from Massachusetts to Puerto Rico, and then back to the continental United States before deportation to Ecuador. Others paths show migrants being shuffled back and forth between the same facilities or geographically proximate facilities, as was Oscar's (Figure 5.5) experience. While Franklin was in the same place for all 16 weeks of his detention, Clemente passed through five detention places in just four weeks.

These often baffling movements through space and time can be partially attributed to a mismatch of available facilities with places of capture. A 2009 report by the Transactional Records Access Clearinghouse at Syracuse University (TRAC 2009) states,

7 The 'interior' means far from land borders crossed to enter the United States without a visa. As shown in the maps, for Ecuadorians the 'interior' often references the east coast, due to historic settlement patterns.

> As the number of detainees has grown, the agency [ICE] – at least until recently – has not sought to balance where it located new detention beds with where the individuals were apprehended. Instead ICE has adopted a free-wheeling transfer policy to deal with the resulting imbalances. Under this policy, ICE transports detainees from their point of initial ICE detention to many different locations – often over long distances and frequently to remote locations.

A lack of planning and coordination within the D & D system is therefore a contributor to the high number of transfers.

Some transfers are plainly orchestrated to curtail migrants' rights (Morawetz 2005, HRW 2009). As Martin (2012: 329) writes, 'Transferring detainees between facilities allows ICE to manipulate 'due process', exploiting the US Federal Court's uneven legal geography to attain its desired legal outcomes'. For example, Kanstroom (2007b) reported that immigrants apprehended during a 2007 raid in New Bedford, Massachusetts, were quickly transferred to detention facilities in Texas before they could inform family members and meet with lawyers. What's more, the rate of transfers to court districts known to be hostile to detainees' claims are notably higher than to those that are not (Morawetz 2005, HRW 2009). Migrants are also frequently transferred to places where no pro bono lawyers are available (Fernandes 2007, HRW 2010, ACLU 2011). Furthermore, mobility can be used to discipline and control detainees (Gill 2009). For example, detainees have been transferred in response to protests regarding treatment and facility conditions (Bernstein 2010b), and known organizers have been moved to pre-empt continued organizing (Dow 2004).

Profit generation also plays into these chaotic space-time geographies. The fact that a detention centre makes money for every day it houses a migrant contributes to extended detention stays because managers may not be motivated to facilitate a detainee's departure. That per day fee also influences migrants' circuitous transfer paths, as different players wrangle to fill 'beds' (Dow 2004, Barry 2010, Cervantes-Gauschi 2010, Wood 2011). Thus, while multiple transfers and illogical routes occur due to a lack of planning and central coordination within the D & D system, they also symbolize the power and influence of the immigration industrial complex.

Consequences of Detainee Transfers

Whether due to poor planning or design, these patterns of forced mobility profoundly shape migrants' experiences in the D & D system. Transfers are profoundly detrimental to case outcomes (Morawetz 2005, HRW 2009, NIJC 2010, ACLU 2011). For migrants detained while living in the US interior, transfers usually move them away from their home area, thereby truncating access to familial and legal support networks that could assist efforts to avoid removal (NIJC 2010, ACLU 2011). Transfers interfere with established lawyer-client relationships, problematize detainees' right to select their own counsel,

and change the set of laws applied to particular cases (HRW 2009). Additionally, many detention facilities are located in remote, hard to access areas far from urban areas (HRW 2009; NIJC 2010, ACLU 2011). In fact, as already mentioned, to the DHS, difficulty of access is a positive trait to be sought when selecting detention sites (Martin 2012, cf. Mountz 2010 on Canada). Furthermore, Gill (2009) found that in the United Kingdom, frequent transfers have a disciplinary effect on both facility managers and support groups poised to assist asylum seekers, effectively discouraging them from actually providing assistance. Moran, et al. (2012) argue that circuitous and disorienting transportation routes work to establish the power of the carceral system over prisoners in Russia. In the US detention system, too, transfers make the involvement of assistance entities difficult and undermine the development of relationships with facility personnel in ways that further limit aid available to detainees.

Transfers also contribute to detainees' isolation by further complicating their communication efforts. Migrants, family members and others in support networks often scramble to determine a new location and re-establish communication after a transfer. Adding to this challenge is the fact that calling cards are specific to the facility in which they are purchased, which means that if a detainee is suddenly transferred the card becomes useless. José Carlos explained,

> I had money, but the problem was that sometimes it took four or five days for them to give you the card. You fill out a request for the card and it comes after five days. They came to give me the card and that same day that the card arrived, that day they transferred me and that card was only good there, in the other one it was no good.

Many interviewees reported that their account money arrived at their new location weeks after they did. Until that money arrived they could not purchase a calling card and were thus unable to inform people of their new location.

Many deportees complained about the physical conditions they experienced during transfers, during which detainees are routinely shackled hands and feet to waist (HRW 2009, ACLU 2011). While some interviewees were mildly perturbed by this practice, it was traumatizing for others. Oscar remarked, 'I want to forget that. One of the worst things in my life.' Transfers typically entail lengthy processes for checking migrants out of one facility and into another. When large numbers of migrants arrive at the same time, detainees may have to wait for extended periods, still in chains. Julio reported 40 hours in cuffs during one transfer, and noted that personnel checking in new arrivals took leisurely breaks. Interviewees also told of waiting in crowded rooms without sufficient places to lie or sit and in exceedingly cold rooms, and receiving little food. Julio recalled one miserable transfer experience: 'Because we arrived at 8 p.m. it was too late to check us in, so we had to spend the night in a very cold room without food. Here I really suffered for the cold. It was a cement floor we had to sleep on with no beds or blankets.'

Additionally, many people find the mobility between facilities disorienting and upsetting (Gill 2009, HRW 2009, ACLU 2011). Within the D & D system, transfers are treated as guarded secrets. In its 2009 report, TRAC notes how its own efforts to compile information were complicated by ICE's and contracted facilities' reluctance and refusal to share even basic data regarding transfers. As already explained, some interviewees did not know names or locations of the facilities through which they passed. Most interviewees reported little to no advance notice regarding when a transfer was to occur, to where, or why. Many reported being awoken during the night. Remigio recounted, 'They just arrive at your cell saying 'you are going, get your things' and nothing else. They don't tell you ahead of time or don't let you call your family to let them know you are going to be transferred, none of that.' José Carlos explained how guards sustained a sense of secrecy throughout detainee movements: 'They always moved me chained. They take you in a van … and they enter [detention facilities] by tunnels … you don't even know where you are.' Some interviewees received so little information that during transfers internal to the US they were under the impression they were finally being deported. The realization that the movement was just to another facility was profoundly stressful. When the actual deportation did occur, it was common to be awoken in the night, herded into a room with dozens of others, and held there for hours. Interviewees who flew via JPATS (as opposed to commercial flights) typically received minimal food and were shackled hands, feet and waist.[8]

What's more, transfers contribute to extended detention durations (ACLU 2011). Transfers can further complicate the procedural trajectory, as pieces of information are lost, and paperwork trails (such as for travel documents) are re-routed. A new set of personnel must become familiar enough with the migrant's case to take appropriate procedural steps. In this milieu, different facilities have different procedures and backlogs. This is exemplified in the case of two brothers travelling together, Andrés and Juan, who were both captured by ICE on the Texas border and detained in the Willacy Detention Center in Harlingen, Texas. Juan was transferred to the South Texas Detention Center in San Antonio, due to, as explained to me by an ICE official, lack of bed space in Harlingen. This move led to a much longer detention time for Juan; Andrés returned to Ecuador after three weeks, while Juan returned almost two months later. This example also points to ways in which chance and unpredictability are woven into the fabric of the D & D system.

The cumulative architecture of the D & D system makes detained migrants feel like they are in a purgatory of waiting, a feeling with profound implications. As José Carlos stated, 'Being locked up there you get desperate, because there was no way to communicate with anyone or know when you are going to be sent, nothing.' Daniela, detained two months, watched as a group of Ecuadorians came in after her and stayed only three weeks; she was depressed and frustrated when

8 Of my 40 interviewees, 31 were deported on JPATS flights.

she received no explanation for her longer wait time. Carlos also described feeling desperate when he could not obtain information, and said: 'you suffocate because you don't know when they are going to get you out'. In these accounts, we see how turbulent the D & D system can feel for those caught within it, or how they experience what Welch (2002: 109) calls the 'Kafkaesque nature of detention'.[9] Importantly, experiences of despair, frustration and stress while in detention can influence detainees' decisions to fight deportation (Martin 2012, Coutin 2010, ACLU 2011). Multiple interviewees indicated that they abandoned attempts to gain 'relief' from 'removal' simply because they could not bear to be detained for the time required. Both Saira and Sergio were mentioned earlier as experiencing the longest detention durations of my interviewees. Sergio had a US citizen wife and child, met the requirements for 'Adjustment of Status', and had a private lawyer paid by his employer. After 28 weeks in detention, however, he gave up his appeal because the process would have taken at least another year. Similarly, Saira, who applied for asylum based on claims of domestic abuse in Ecuador, said that after being detained three months she gave up, 'I couldn't stand to be there any more time, locked up...there I was sick all the time, sometimes I could eat and sometimes I couldn't eat, and I couldn't stand to be shut in anymore, so I signed my deportation.' Indeed, despite and because of these 'Kafkaesque' conditions, most detainees eventually become deportees.

Conclusion

While there is no central, sinister orchestrator behind the chaotic geographies mapped in this chapter, aspects of the chaos that are clearly calculated dangerously interlock with those that may not be, with significant consequences. The conceptualization of detention as a source of revenue contributes to a continuous push for policies to detain and deport increasing numbers of people. The ever-higher deportation target numbers have led to overcrowded facilities, harried employees and a clashing multitude of organizational structures. Systemic detainee mistreatment and rights violations are linked to profit optimization. The resulting structure of the D & D system creates unpredictability and confusion with important consequences for detainees, including nearly guaranteeing their eventual deportation. Rhetoric pushed by media pundits and politicians dehumanizes immigrants in ways that distract the public from concerning themselves with these realities. Finally, the disorder works to conceal the powerful capitalist forces driving detention expansion, as well as the waste and inefficiency characteristic of the resulting detention apparatus.

9 This references a 1937 play by Franz Kafka that tells the story of a man trying to defend himself without knowing the charges and evidence against him or how long his imprisonment and trial will last.

Because factors behind these chaotic geographies are multiple and overlapping, the project of changing or dismantling these geographies may seem daunting and overwhelming. I venture, however, that the chaos actually offers numerous cracks through which change may be affected. That is, there are many possible tactics for disrupting existing geographies, and our strategies can and must be multi-faceted. For example, activists can push for regulation and standardization of detention centre conditions, work to increase the information and legal assistance available to detainees, and draw attention to the taxpayer dollars wasted due to inefficiency and lack of planning. In addition, migrant support organizations can adopt strategies to match the existing chaos, such as developing dispersed, localized and yet highly integrated networks across detention sites. In addition, it is critical that we remember the goal is not to bring order to the chaos to just build a 'better' D & D system. Instead, those working for change must also target the driving forces behind the existence of the system. We must, therefore, work for broader structural changes regarding the economic dependencies that have been created, the intimate ties between corporations and policymakers, and the popularity of anti-immigrant discourse.

This study also illustrates the importance of empirical work across transnational settings. By conducting research on US practices from a country of migrant origin, we gain a rich and lucid understanding of detained migrants' experiences and implications of these experiences. Such a methodological approach in itself suggests possible strategies for altering these chaotic geographies. For example, the dearth of information regarding transfer practices may be addressed by essentially circumventing the ICE to collect data directly from deportees. Such tactics point to cooperation across borders, between support organizations in the US and in countries of migrant origin. What's more, approaches incorporating countries of migrant origin provide opportunities to bring additional, previously unheard voices into the conversation.

References

ACLU. 2011. Outsourcing responsibility: The human cost of privatized immigration detention in Otero County. http://aclu-nm.org/wp-content/uploads/2011/01/OCPC-Report.pdf [last accessed 20 July 2011].

Acosta, A., S. López and D. Villamar. 2004. Ecuador: Oportunidades y amenazas económicas de la emigración. In *Migraciones: Un juego con cartas marcadas*, ed. F. Hidalgo. Quito: Ediciones Abya-Yala, 259-301.

American Friends Service Committee (AFSC) 2010. Locked up but not forgotten. http://afsc.org/sites/afsc.civicactions.net/files/documents/LockedUpFINAL.pdf [last accessed 20 July 2011].

Barry, T. 2009a. A death in Texas: Profits, poverty, and immigration converge. *CIP Americas*, posted October 26, 2009. http://www.cipamericas.org/archives/1894 [last accessed 20 July 2011].

Barry, T. 2009b. National security business on the border. *Americas Program*, posted September 27. http://www.cipamericas.org/archives/1858 [last accessed 20 July 2011].

Barry, T. 2010. The shadow prison industry and its government enablers. *CIP Americas*, posted January 29. http://www.cipamericas.org/archives/1995 [last accessed 20 July 2011].

Bernstein, N. 2010a. Officials hid truth of immigrant deaths in jail. *The New York Times*, 9 April. http://www.nytimes.com/2010/01/10/us/10detain.html?_r=1&ref=incustodydeaths [last accessed 20 July 2011].

Bernstein, N. 2010b. Volunteers report on treatment of immigrant detainees. *New York Times*, 29 April. http://www.nytimes.com/2010/04/29/nyregion/29visitors.html?_r=1&sq=Volunteers%20Report%20on%20Treatment%20of%20Immigrant%20Detainees [last accessed 20 July 2011].

Bloch, A. and L. Schuster. 2005. At the extremes of exclusion: Deportation, detention and dispersal. *Ethnic and Racial Studies* 28(3): 491-512.

Cervantes-Gautschi, P. 2010. Wall Street and the criminalization of immigrants. *Americas Program*, 10 June. http://www.cipamericas.org/archives/3304 [last accessed 4 August 2011].

'Chaos'. 2001. In *Random House Webster's College Dictionary*. New York: Random House, 222.

Coleman, M. 2007a. A geopolitics of engagement: Neoliberalism, the war on terrorism, and the reconfiguration of US immigration enforcement. *Geopolitics* 12: 607-634.

Coleman, M. 2007b. Immigration geopolitics beyond the Mexico-US border. *Antipode* 39 (1): 54-76.

Coleman, M. 2009. What counts as the politics and practice of security, and where? Devolution and immigrant insecurity after 9/11. *Annals of the Association of American Geographers* 99 (5): 904-913.

Collyer, M. 2007. In-between places: Trans-Saharan transit migrants in Morocco and the fragmented journey to Europe. *Antipode* 39 (4): 668-690.

Coutin, S.B. 2010. Confined within: National territories as zones of confinement. *Political Geography* 29: 200-208.

Davis, A.Y. 2003. *Are prisons obsolete?* New York: Seven Stories Press.

Detention Watch Network. 2010. About the US detention and deportation system. http://www.detentionwatchnetwork.org/aboutdetention [last accessed 20 July 2011].

DHS Office of Immigration Statistics. 2011. Immigration enforcement actions: 2010. http://www.dhs.gov/xlibrary/assets/statistics/publications/enforcement-ar-2010.pdf [last accessed 24 July 2011].

Dow, M. 2004. *American gulag: Inside U.S. immigration prisons*. Berkeley and Los Angeles: University of California Press.

Fernandes, D. 2007. *Targeted: Homeland security and the business of immigration*. New York: Seven Stories Press.

FLACSO. 2008. *Ecuador: La migración internactional en cifras*. Quito, Ecuador: United Nations Population Fund and FLACSO.
Flynn, M. and C. Cannon. 2009. *The privatization of immigration detention: Towards a global view*. Geneva: Global Detention Project.
Gerlach, A. 2003. *Indians, oil, and politics: A recent history of Ecuador*. Wilmington, DE: Scholarly Resources, Inc.
Gill, N. 2009. Governmental mobility: The power effects of the movement of detained asylum seekers around Britain's detention estate. *Political Geography* 28: 186-196.
Golash-Boza, T. 2009a. A confluence of interests in immigration enforcement: How politicians, the media and corporations profit from immigration policies destined to fail. *Sociology Compass* 3 (2): 283-294.
Golash-Boza, T. 2009b. The immigration industrial complex: Why we enforce immigration policies destined to fail. *Sociology Compass* 3 (2): 295-309.
Gratton, B. 2007. Ecuadorians in the United States and Spain: History, gender and niche formation. *Journal of Ethnic and Migration Studies* 33(4): 581-599.
Hernández, D.M. 2008. Pursuant to deportation: Latinos and immigrant detention. *Latino Studies* 6: 35-63.
Human Rights Watch. 2009. Locked up far away: The transfer of immigrants to remote detention centers in the United States. http://www.hrw.org/en/node/86789 [last accessed 20 July 2011].
Hyndman, J. and A. Mountz. 2008. Another brick in the wall? Neo-refoulement and the externalization of asylum in Australia and Europe. *Government and Opposition* 43 (2): 249-269.
Jokisch, B. and D. Kyle. 2008. Las transformaciones de la migración transnacional del Ecuador, 1993-2003. In *La migración ecuatoriana: Transnacionalismo, redes e identidades*, edited by G. Herrera, M. C. Carrillo and A. Torres. Quito: FLACSO, 57-69.
Kanstroom, D. 2007a. *Deportation nation: Outsiders in American history*. Cambridge, MA: Harvard University Press.
Kanstroom, D. 2007b. Reaping the harvest: The long, complicated, crucial rhetorical struggle over deportation. *Connecticut Law Review* 39 (5): 1911-1922.
Kyle, D. and B. Jokisch 2005. Leaving Ecuador. *Berkeley Review of Latin American Studies* Spring: 45-48.
Martin, L. 2012. 'Catch and remove': Detention, deterrence, and discipline in US Noncitizen family detention practice. *Geopolitics* 17 (2): 312-344.
Martin, L.L. and M.L. Mitchelson. 2009. Geographies of detention and imprisonment: Interrogating spatial practices of confinement, discipline, law, and state power. *Geography Compass* 3 (1): 459-477.
Miller, T.A. 2003. Citizenship & severity: Recent immigration reforms and the new penology. *Georgetown Immigration Law Journal* 17 (4): 611-666.

Moran, D., L. Piacentini and J. Pallot. 2012 Disciplined mobility and carceral geography: Prisoner transport in Russia. *Transactions of the Institute of British Geographers* 37, 3, 446-460.

Morawetz, N. 2005. Detention decisions and access to habeas corpus for immigrants facing deportation. *Boston College Third World Law Journal* 25: 13-33.

Mountz, A. 2004. Embodying the nation-state: Canada's response to human smuggling. *Political Geography* 23 (3): 323-345.

Mountz, A. 2010. *Seeking asylum: Human smuggling and bureaucracy at the border*. Minneapolis: University of Minnesota Press.

National Immigrant Justice Center. 2010. Isolated in detention: Limited access to legal counsel in immigration detention facilities jeopardizes a fair day in court. http://www.immigrantjustice.org/isolatedindetention [last accessed 20 July 2011].

Preston, J. 2011. U.S. Pledges to Raise Deportation Threshold. *New York Times* 17 June: http://www.nytimes.com/2011/06/18/us/18immig.html?_r=1, [last accessed 8 August 2012].

Sullivan, L. 2010. Prison economics help drive Arizona immigration law. *National Public Radio*, 28 October: http://www.npr.org/templates/story/story.php?storyId=130833741 [last accessed 26 July 2011].

TRAC. 2009. Huge increase in transfers of ICE detainees, *TRAC Reports.* http://trac.syr.edu/immigration/reports/220/ [last accessed 20 July 2011].

TRAC. 2010. Backlog in immigration cases continues to climb, *TRAC Reports.* http://trac.syr.edu/immigration/reports/225/ [last accessed 20 July 2011].

TRAC. 2011. New judge hiring fails to stem rising immigration case backlog, *TRAC Reports.* http://trac.syr.edu/immigration/reports/250/ [last accessed 20 July 2011].

Weizman, E. 2007. *Hollow land: Israel's architecture of occupation*. New York: Verso.

Welch, M. 2002. *Detained: Immigration laws and the expanding I.N.S. jail complex*. Philadelphia: Temple University Press.

Wood, G. 2011. A boom behind bars. *Business Week*, March 17. http://www.businessweek.com/magazine/content/11_13/b4221076266454.htm [last accessed 20 July 2011].

Chapter 6
Up the River (from Home): Where Does the Prisoner 'Count' on Census Day?

Matthew L. Mitchelson

Introduction

A political controversy has emerged in the United States at the intersection of two spatial processes: census-taking and imprisonment. This chapter analyses that controversy through the combined lens of census- and carceral-geographic research, and fills an empirical gap in the controversy's scholarly and legal discourse by adding the interviewed voices of formerly imprisoned people to the conversation. The paper consists of four sections. In the following section I analyse the controversy using: first, a Foucauldian approach to census-taking as the production of 'spatial legibility' and 'calculable territory' (Hannah 2000, 2009, Rose-Redwood 2006, 2008); and, second, carceral-geographic work that insistently demonstrates the *simultaneous* fixity and mobility produced and sustained via imprisonment (Gilmore 2007, 2008, Moran et al. 2012). Section three is focused on formerly imprisoned participants' reactions to the themes detailed in the previous section, and to the census controversy itself. The fourth, final section offers a discussion of what is at stake in the controversy's origins and its possible outcomes through the lens of 'statistical citizenship' (Hannah 2001). In the remainder of this introductory section I outline the controversy and describe the methodological approach employed for this research.

Prisoners and the Census: A Controversial Count

As Lotke and Wagner note, 'an accurate count of the population is so fundamental to representative democracy that the framers of the Constitution required it in the opening paragraphs' (2005: 588). The decennial census is distinct from other social surveys in the USA because it is used as the basis for the nation's political districts at a variety of scales. Its integrity hinges on the act of counting each resident 'once, only once, and in the right place' (National Research Council 2006). Prisoners are currently tabulated as residents of the prison in which they are held (U.S. Census Bureau 2006). However, most prisoners have come from cities far away from the prison

in which they are confined and prisoners generally cannot vote (Lotke and Wagner 2005, Ho 2011). Prisoners therefore appear as 'phantom constituents' on paper when prison populations are included in census counts that are used for political purposes from which prisoners are generally excluded (New York Times Editorial Board 2006).

The impacts of these counts in a representative democracy are generally internalized by states because the majority of prisoners do not cross state lines, which would impact the distribution of US Congressional seats (i.e., inter-state reapportionment). Instead, 'counting' prisoners as residents of prisons directly impacts intra-state, inter-electoral district 'proportionality' (i.e., 'one person, one vote'), and can therefore be interpreted as an example of malapportionment, or, 'inequality of population among electoral districts' (Johnston and Pattie 2003, Morrill 1981: 10). For example, during its 2002 election Anamosa, Iowa, was divided into four City Council Wards of roughly equal size (ranging from 1,366 to 1,377 residents). However, more than 95 per cent of Ward 2's residents were confined in the Anamosa State Penitentiary. In turn, 58 'actual constituents' were thus granted the same level of political representation that was divided amongst no fewer than 1,366 residents of the other Anamosa Wards; or, Ward 2 constituents were apportioned the electoral power of roughly 25 voters in other wards (Roberts 2008, Ho 2011).

The vastly disproportionate incarceration of people of colour in the United States further complicates the already intractable question of race-informed political representation under the country's Voting Rights Act and related legislation (Tonry 1995, Gilmore 2007, Wacquant 2001, Forest 2001). Ultimately, the seemingly innocuous act of counting prisoners reconfigures both the demography of the communities from which prisoners come and the communities to which prisoners go, as prisoners' move through the criminal justice system (Lotke and Wagner 2005, Levingston and Muller 2006, Stinebrickner-Kauffman 2004). Prisoners' mobility thus becomes an extremely important object of political-geographic inquiry as the choice of how – which is to say where – to count prisoners can impact places and census geographies across the carceral landscape (Ho 2011, Kelly forthcoming).

The question of where prisoners should count during the US decennial census is predictably controversial (Webster 2004, Roberts 2010). The U.S. Census Bureau (2006) primarily defends their practice in terms of cost and operational efficiency. Critics of the practice have called it 'the 5/5ths Clause' and insist that it enables 'prison-based gerrymandering, and violates the principle of 'one person, one vote' (Marantz 2006, Prison Policy Initiative 2011). Congressional hearings were held before and after Census 2010 (Wagner et al. 2006, Groves 2010, Wagner 2010). The controversy remains largely unresolved in the wake of the most recent Census Day, but the Bureau has provided a technical fix by releasing prison population figures at the block level for redistricting purposes (Groves 2010). A few states have compiled prisoners' pre-incarceration addresses and are reallocating prisoner counts

(Prisoners of the Census 2012). In other places state and local governments will have the option to work around the important problem of malapportionment, but, they will generally do so by removing the prison population altogether. The question of where prisoners should count will be moot. Prisoners will not count at all.

Methodological Approach

Prisoners are at the heart of this particular census controversy, yet few prisoners are visibly involved in the debate. This work argues that former prisoners are uniquely positioned to comment on the census controversy – to explain both how it has come to be and what the stakes are for those being counted. I interviewed former prisoners as part of my research on prisoner location and the decennial census and this paper reports the preliminary findings of those interviews. I asked people who had been through various prison systems *where* they might want to be counted and *why*. In response to Hannah's call for 'a more 'nuts and bolts' political explanation' of the ways that numbers and counting both register and transform reality (Hannah 2001: 517), I also wanted to understand *how* these particular census counts come to be from a very important vantage point: the prisoner's.

I must acknowledge that practical matters of timing (I completed this work under the limited research schedule of a graduate student) and limited access to Department of Corrections facilities constrained the participant population in at least one significant manner. Ideally this research would have included interviews with those enumerating and enumerated within prisons on Census Day: those for whom the political stakes of the census controversy are most explicitly linked. However, formerly incarcerated participants were able to reflect on their imprisonment as well as their pre- and post-incarceration lives. Importantly, former prisoners have been all the way 'through' the spatial process of imprisonment and can speak to this holistic process entailing arrest, adjudication, prison admission, displacement, captivity, institutional transfers, release and re-entry.

I conducted 50 semi-structured interviews with former prisoners. All interviews took place in Athens, Georgia, in 2008-2009. The Athens Justice Project (AJP), a local non-profit organization and I formed an indispensable research partnership during this process. The AJP office is conveniently located and provided a locally recognizable and comfortable setting for participants and it functioned as an ideal host site for these interviews (Elwood and Martin 2000). Roughly one third of the participants were introduced by Athens Justice Project employees, interns and clients. Roughly two thirds of the participants were snowballed from this AJP-related population, locally distributed recruitment fliers and informal word of mouth. Demographic characteristics were self-reported by participants and generally mirrored state-wide trends in that

participants were most frequently men of colour. The average participant was in their 40s, which is older than both the state's prison and total populations, and reflects the fact that participants had been through the prison system (aging throughout the process).

I intended to learn what participants could teach me about existing carceral-geographic research. I also needed to learn what I didn't know – what I couldn't know without having been imprisoned myself, and what I couldn't learn from a book or a dataset (Richards and Ross 2001). I thus employed a semi-structured method of interviewing, which makes use of an interview protocol, but also allows participants to actively shape the interview (Dunn 2005). I used an unobtrusive digital voice recorder during the interviews in 49 of the interviews. In one instance, a participant requested that I not record the interview. I transcribed the interviews and triangulated them with official Georgia Department of Corrections records and my personal field notes and research journal (Janesick 1999, Hodder 2000, Janesick 2000, Cresswell 2003). This data forms the heart of the research reported in section three of this chapter. Before reporting findings from this work, I survey geographies of the prison, the census, and their intersection.

The Controversy in the Context of Census- and Carceral-Geographic Research

Carceral Geographies and the Spatial Structure of Imprisonment

Lauren Martin and I (2009) have argued that imprisonment is characterized by the tension between two spatial forces: fixity and forced mobility. On the one hand imprisonment clearly emplaces human beings, at times to the point of complete immobilization. However, at the same time, it remains true that human beings are not born into imprisonment – every prisoner has made the (albeit coerced) migration 'up the river' to a prison from their pre-arrest 'home,' which is understood to entail a physical location and, at the same time, 'extend far beyond a material dwelling' (Blunt and Dowling 2006: 13). The spatial structure of imprisonment can thus be thought of as entailing origins and destinations in a process of incarceration-as-migration.

Our work resonates strongly with that of Moran et al. whose research on 'etap' in the Russian prison system demonstrates the 'movement to and from, or between' the institutions that make up various prison systems (2012: 4). As several contributors to this volume (in particular see chapters by Moran et al, Hiemstra, Martin and Gill) make clear, the seeming fixity of detention and imprisonment must be actively produced and maintained, often via governmental techniques that institutionalize mobility and population movements. There is a crucial, often overlooked, politics of mobility that impacts both the immobilized

and their immobilizers in carceral contexts (Gill 2009). As Gilmore and Gilmore (2008: 144) explain:

> The connection between the rise of the nation-state and the rise of the prison is located in the contradiction between mobility and immobility: when the conditions attending on a global system that requires constant motion (i.e., capitalism) clash with challenges to maintain order, spatial fixes such as racialization and criminalization temporarily settle things through complicating insider-outsider distinctions with additional, rights-differentiated hierarchical schemes.

The seemingly contradictory spatial form of imprisonment simultaneously forms and reflects these social contradictions (Soja 1980). In effect, the geographically fragmented lives of prisoners serve as a spatial fix for geographically fragmented states, which are both relentlessly local and 'nonlocalizable' (Gupta 1995: 390) at once, and of which Solzhenitsyn's 'archipelago' is perhaps emblematic. I will often contrast two vantage points in what follows: the prisoner's and the state's. However, both 'the' prisoner and 'the' state must be noted as geographic representations of large, diverse and often inconsistent populations and institutions (i.e., prisoners and states).

As participants in my research explained (see below), the contradictory forces of fixity and mobility produce a dyadic geography in prisoners' lives: physical presence and (most often) socio-political exclusions characterize the site of their imprisonment; physical absence and increasingly strained social ties characterize the place from which they have come. Ruthie Gilmore (2008) has conceptualized these spatially discontinuous places as a singular geographic unit, which she calls an 'abandoned region'. From a prisoner's vantage point in particular, the abandoned region has much to offer as a unit of analysis, in terms of analytical accuracy. This conceptualization also performs an ontological reclamation of the prison (i.e., an architectural fact) as an immanently human geography (cf. McKittrick 2012, Gilmore 2007). But the structure of imprisonment dictates that the abandoned region is necessarily a geography of human life on the move. As Starkweather's (2009) research demonstrates, life on the move can be enumerated in multiple ways – through 'objective' calculating techniques, or, through affective nation-building techniques of inclusion/exclusion, for example (see below). Before turning to the affective geographies of imprisonment and census taking in what follows, I turn now to the political problem of spatially accounting for a population on the move within a relatively static census geography.

Geographies of Census-Taking and Spatial Legibility

Geographers are constructing a vibrant critical geography of census taking and other geo-coding projects (Hannah 2000, 2001, Rose-Redwood 2006, 2008, Starkweather 2009). These explicitly spatial analyses offer welcome contributions

to recent critical scholarship concerned with the relationship between numbers and political power (Rose 1999: 197-232, Cohen 1999: 150-204, Nobles 2000). Starkweather's research on the enumeration of overseas populations during the decennial census – particularly her analysis of nation-building, citizenship and belonging – is particularly instructive alongside participant responses (see below), and here I focus primarily on the first two scholars. Hannah's (2000) research is empirically based on census taking during the nineteenth century – when the country's prison population was a fraction of its current size – yet his analysis sheds light on the current controversy. Hannah argues that state power in the context of census taking requires 'a complicated spatial logic' (2000:13) capable of balancing contradictory processes of fixity and flow – the same spatial characteristics that structure imprisonment. However, from the vantage point of state power, 'the most legible population is one perfectly assorted into nuclear families, each occupying a single, detached home at a fixed address' (Hannah 2000: 125). The lives of individual prisoners and their families present a stark contrast to this standard, as incarceration-as-migration sets lives in motion and spatially fragments households, which are understood by Gilmore as 'fractured collectivities' (2007: 5) in the context of imprisonment. She explains that where the prison is concerned, 'households stretch from neighbourhood to visiting room to courtroom, with a consequent thinning of financial and emotional resources' (Gilmore 2007: 16).

Like Hannah, Rose-Redwood's (2006: 471, 2008) critical geo-coding research suggests that 'state projects of legibility and simplification' (after Scott 1998) are always fundamentally geographic affairs. Legibility marks territory through a system of spatial references and renders territory both readable and physically accessible to the state (Hannah 2000: 14-17, 2009). But in the case of prisoners and the census we find 'the' state (i.e., multiple governmental entities and jurisdictions) at cross purposes. The sovereign power to seize bodies, and the disciplinary-spatial power relations that set the population geographies of imprisonment into motion ultimately contradict representations of prisoners as the spatially fixed objects of governmentality – the type of spatial legibility – that a census requires. As Hannah explains, 'acts of observation can be improved not merely by engaging experts to do the looking, but by concentrating on the most legible units' (2000: 133).

The geography of imprisonment is structured by a carceral archipelago, fractured households, and population flows between these types of places across abandoned regions. If the census is to produce a geographic framework that exhausts the territory of the United States without overlap, it must choose between individual prisoners and prisons (or, more accurately, prison populations). From the state's vantage point(s), to observe and enumerate the abandoned regions of imprisonment requires an investment in the bodies and lives of individual prisoners – an impossibility, since these regions are defined by abandonment. Prisons – fixed, immovable architectural facts – are far more legible to the state in this context. Most of the time (i.e., 56.3 per cent of the time in Census 2000)

administrative records were used as the source for Individual Census Reports (U.S. Census Bureau 2006). These would generally be the Census Day roll call tally, an average daily population for that institution, or the institution's rated capacity. In other words, prisoners may not count towards current redistricting, but most of them were never directly counted in the first place.

Geographies of Prisoners and the Census

Prisoners, prisons and the prison population must all be integrated as 'calculable territory' by the census; this is a pressing political problem of governance that Hannah (2009) has explained through his model and general framework of that concept (and see Elden 2007). How, which is also to say *where* in this case, do people, places and things 'count' when they have been territorialized in very different ways, and for very different purposes? The question can be asked in both empirical and normative terms. Where do prisoners count? Where should prisoners count? The practice of counting prisoners during the decennial census presently imposes both of these important questions.

The census controversy has been the explicit focus of at least two geographic projects. However, the two projects are of a very different nature. Webster (2004) was the first geographer to note the controversy at hand. He has argued that the US correctional system adversely impacts political representation through both felony disenfranchisement laws and the decennial census enumeration of prisoners detailed above. That argument is part of a larger call to explore questions of social justice through 'equity-oriented pedagogy' – particularly the political power of minority groups in systems based on proportional representation. Webster concludes that current enumeration procedures can create unjust levels of representation in both jurisdictions from which prisoners have come (i.e., where political power can be decreased) and to which they go (i.e., where political power can be increased).

Burayidi and Coulibaly report their efforts – initiated on behalf of a Wisconsin city administrator and the president of an economic development corporation – 'to present a 'true' profile' of communities that host prisons (2009: 141). Like Webster, the authors make note of the political impacts of prisoner enumeration. However, their primary concern is that 'communities with large prison populations may be portrayed less favourably in Census Bureau reports than their actual socioeconomic profiles may dictate' (Burayidi and Coulibaly 2009: 148). Most of the article details a procedure for recalibrating data from different sources and with different underlying spatial units so as to remove correctional populations from local profiles. The authors find, among other things, that 'the actual number of residents in the city (7,042) is smaller than that reported by the Census Bureau (11,485). Also, the proportion of the White population in the city increases from 85.6 per cent to 96.5 per cent' (ibid.: 147).

The Burayidi and Coulibaly project ultimately appears to observe the same underlying socio-spatial processes explained by Webster, but from (perhaps

diametrically opposed) vantage points and/or questions of social justice. My aim, in what follows, is to engage both projects directly, as well as the other research surveyed in this section, with the experiences and voices of prisoners themselves.

The Controversy from the Vantage Point of former Prisoners

The Carceral Geography of Participants' Life Experiences

The spatial structure of imprisonment is reflected in the life histories and geographies narrated by participants in my research. Participants were all residents of Athens-Clarke County (where I worked at the time) or an adjacent county. However, the experiences of former prisoners make up the research study area. In turn, Athens-Clarke County was observed to be part of a large and geographically complex abandoned region. In sum, participants had served a total of 59,195 days (i.e., just over 162 years) in Georgia's state prisons. Participants' experiences in the Georgia prison system ranged from one, relatively short imprisonment (90 days), to seven imprisonments totalling more than 20 years.

Participants' incarceration histories also involved an expansive geography of imprisonment. Most participants (26) had served only one sentence in a Georgia state prison; 18 participants had been imprisoned multiple times; seven participants had been imprisoned five times or more. All of the participants save one had served time in local jails. All but two of the participants served time in state prisons: 45 participants served in Georgia; in addition to Georgia, participants served time in Florida, Kentucky, New York, Maryland and Virginia state prisons. One participant was transferred between 11 institutions in 8 states while serving a single sentence under the Federal Bureau of Prisons' jurisdiction.

'They Shipped Me': Narrating Incarceration-as-Migration

Two sets of themes emerged during the interviews, both of which are directly related to the spatial structure of imprisonment. The first set of themes reflects a sense of spatial instability during processes of forced migration. The commonplace experience of being admitted and/or transferred – taken into and/or moved across the carceral archipelago – was most commonly spoken of as either 'shipped' or 'housed out'. Neither term was part of the interview protocol, yet these specific terms were spoken (unprompted) by participants in the majority of interviews (i.e., in 37 cases). When K.H. and I spoke, he had served a total of four and a half years (more than 20 per cent of his adult life) in prison for several non-violent crimes. Much of our interview centred on the impacts of incarceration on his relationship with his wife and son, which included the burdens imposed by his separation from them and compounded by physical distance and frequent relocation. When I asked K.H. to recount the places prisoners go after sentencing he explained:

> It's different, really, for each individual. The first time [I went to prison] I ended up at 10 or 12 different places. Once you receive your sentence, you go back to the county jail and wait for a bed to come open in a prison, and the county houses you. Walton County [where K.H. was sentenced] was full, and there's a constant turnover, so I was housed out to other counties besides Walton.
>
> I believe it was Walton County first, then I went to Morgan County. I was in Meriwether County, then I was in Lamar County and Dekalb County, then back to Walton County. And that was just in the county jail experience. That was probably in a six-month span.
>
> I pretty much changed institutions once a month. It may have been the situation where I went to Lamar County for a week, and then a bed came open in Meriwether County and they shipped me to Meriwether County for a month. Then I'd go back to Walton. I never knew, when I woke up, if I was going to be there that day.

Some of this population churning results from overcrowding, which has been a substantial problem in the Georgia case for several decades. In the Georgia case, 'buses roll' twice a week, transporting prisoners for admissions and transfers. During related fieldwork at some two dozen state facilities, I was consistently struck by the prevalence of two material artefacts: mile upon mile of concertina wire, and fleets of prisoner transport vehicles; or, signs of fixity and mobility.

The lingering question in all of this is 'to where' (and – at the same time – 'from where') were participants shipped? The distributions of unincarcerated and prison populations in the Georgia case are inversely related. Whereas roughly 70 per cent of the state's population resides on or upland from the piedmont (where Atlanta predominates as a population centre), 30 per cent of the state's population resides in a more or less uniform distribution across the agricultural lands of the state's coastal plains. Conversely, 30 per cent of the state's prison population is held on or upland from the piedmont, and 70 per cent of the state's prisons are located on the coastal plain. In a related study I found that across all prison admissions in the state (1990-2006) the average prisoner was held 100 miles from their pre-arrest residence; in hundreds of cases, the distances exceeds 300 miles (Mitchelson 2012). Again, prisoners are counted as residents of the prisons in which they are held. The controversy surrounding this practice suggests that a normative question lingers: where should prisoners count? In response to this question, I turn to themes that emerged from participant narratives.

'Stand up and Be Counted!': Former Prisoners' Responses to the Census Controversy

Participants and I spent time talking about the U.S. Census Bureau and the decennial census during every interview. Participants' familiarity with the census ranged greatly. Some former prisoners had never heard of (or could not recall what they had heard about) the census or the Census Bureau. At other times participants were intimately familiar with the process and the controversy itself. During each interview I asked participants where they would choose to be counted if they had been given the chance on a previous Census Day, or, if they were given the chance during a hypothetical incarceration:

> If you were in prison, and the Census Bureau counted you, where would you want to be counted: in your prison cell, where you'd come from, or somewhere else?

Participants responded overwhelmingly *against* the practice of being counted in their prison cells. Forty-five participants said they would prefer to be counted in either their pre-arrest home or the address they would be released or paroled to after prison. Two participants were entirely indifferent; and one participant preferred to be counted in the prison.[1]

Many of the respondents were well aware of the political importance of the decennial census. When Eddie and I spoke he was approaching 50 years of age. He had served six prison sentences between 1979 and 1990, for a total of 4.5 years, all on non-violent offenses. The Census controversy was something of which Eddie seemed to be well aware – he was serving time on Census Day (i.e., 1 April) in 1990 during direct enumeration – and he was troubled by the prospect of 'not counting'. Eddie became visibly agitated as he explained:

> When they came to [prison] to do the census a lot of guys were making a joke out of it. They wanted to know if we were working there – you know, stuff like that.
>
> The guys would respond, 'Oh yeah, I'm sweeping and mopping now!' I'd say, 'You know this is a serious thing, this census?' So, then, after the census people had left, some of the people that were making the joke said, 'Are the people still here giving that census?' They wanted to get right with the folks then. But, like the guy with the mop, I said to him, 'This is very important. It's just like Martin Luther king said, stand up and be counted!' Stand up and be counted. So, everything they asked me, I complied. I knew what it was about.

1 These figures tally to 48 because two interviews were so informative that follow-up interviews were arranged. I ultimately conducted 50 interviews with 48 participants.

Participants like Eddie were often not only aware that the census is important, but why (i.e., what is potentially at stake in being counted). For example, Tony had lived much of his life behind bars, both in Georgia and in New York, where he lived as a younger man. My interview with Tony was so informative that we met twice. During our first meeting, Tony explained the overt state racism signalled by the 3/5ths Clause:

> This Census issue reminds me of – back in the days when the North and the South were in conflict about slave issues – the 3/5ths Compromise. Because they wanted the power and the vote, they wanted every 5 slaves to count as 3 white voters. That way they could increase their political power and make laws. They could have the power to control the House and so on.

In Tony's case, as in nearly every other case, 'counting' was viewed in more than mathematical terms.

Starkweather's analysis of overseas enumeration (i.e., 'counting' Americans abroad) is instructive by way of analogy to the intra-national exile produced by incarceration-as-migration. Starkweather (2009: 245) identifies 'two distinct governmental rationalities underlying [census-based] statistical procedures,' the first of which – the production of calculable territory, including population data – has been the primary focus of discourses surrounding the prisoner census count in the United States. However, a second discourse – that of a 'nation-building census' also emerges from Starkweather's work (2009: 242). Inclusion, and, relatedly, exclusion form the basis for many of the quotations Starkweather reports related to being counted during the census: 'it is, first of all, a feeling of belonging to the American nation, of being part of the American people, for instance; or, conversely, feeling 'invisible in the eyes of the Census Bureau' or 'treated as nobodies' (ibid.: 242). Starkweather (2009: 245) concludes that 'the population census expects, envisions, and enacts a static relationship between population and territory that has never been "true" and is only becoming less so'. On the cutting edge of the question posed in this chapter – where does the prisoner 'count'? – rests a normative question of 'home' and its geography.

As was the case in Starkweather's research, participants in my research often identified 'counting' in terms of citizenship. For example, as K.H. explained:

> I would say – I think it would be right – for me to be counted in my pre-incarceration home. That's the most recent place that I paid my taxes, and that I had either a home or a mortgage, or an apartment at least.

Participants repeatedly emphasized (in 41 interviews) that 'a prison is not a home' (cf Libal, Martin and Porter 2012). Waller served 300 days on a forgery charge. Although Waller's sentence had been relatively short (compared to other participants) yet, like all incarcerations, it had a tremendous impact on her life

and her sense of home. For example, when I asked Waller where she'd want to 'count' she explained:

> I never thought of prison as home. I was just doing what I had to do until I could get home. I mean, in my dreams, I would spend my free time trying to focus on what home looked like in my memories. And my parents would send me pictures of home, which I had a whole bunch of different pictures of Athens, and stuff, to remember myself. I could say, 'this is the street I grew up on.' I could stay focused. Because, if I didn't, I would lose my mind. Yeah. Because, when you're in there, you are not a person. You are a statistic.

Under these circumstances, the question of 'home' is ultimately a question of belonging and the (often external) recognition of one's existence. Incarceration – as a migratory process – internalizes both physical and social displacements. Counting the imprisoned population as 'citizens' of a physically present, yet (most often) politically omitted, 'prison nation' appeared to compound participants' experience of these displacements in both a physical and a social sense. This seemingly benign spatial representation – the act of being enumerated – appears to exude a sense of exclusion and domination, or, placelessness.

Discussion

The power to count and account for human beings during the census clearly entails an important spatial politics (Hannah 2000: 113-159, Starkweather 2009). The census controversy of where prisoners should count is exemplary. During this '*unambiguously national* technique of power' (Hannah 2000: 9; emphasis in original), prisons are indexed to the wider census geography of the United States. But as Starkweather (2009: 245) argues, 'the difficulty of re-territorializing the techniques that evolved to govern the Westphalian nation-state' is a pressing concern for governing bodies and scholars alike. Hannah (2001: 516) coined the phrase 'statistical citizenship' and defines it as 'a strategic active participation in the construction of the statistical representations by which individuals are constituted as political actors, objects of social policy, and/or consumers'. He argues the representations produced through censuses and other surveys may be of more political importance than voting, though being counted appears to be a passive form of representation as opposed to casting one's vote.

I want to conclude this chapter by proposing that the census controversy at hand illustrates both the shortcomings and political potential of statistical citizenship for prisoners. The key question – where the prisoner counts – revolves around an individual's capacity to self-consciously and politically manage their statistical representation. In this chapter I have analysed the controversial practice of enumerating prisoners during the decennial census through the lenses of carceral-geographic research and a Foucauldian approach to census taking. Statistical

citizenship in the context of prisoners and the census entails an important set of questions for future research. First, how are the corporeal geographies of prisoners represented during incarceration-as-migration, across various abandoned regions and across various state territories/jurisdictions? And, second, what topological relations connect prisoners, who are in very specific places by definition, to other distant places?

References

Blunt, A. and R. Dowling 2006. *Home*. London and New York: Routledge.

Burayidi, M.A. and M. Coulibaly 2009. Image busters: how prison location distorts the profiles of rural host communities and what can be done about it. *Economic Development Quarterly* 23(2): 141-149.

Cohen, P.C. [1982] 1999. *A Calculating People: The Spread of Numeracy in Early America*. New York and London: Routledge.

Cresswell, J.W. 2003. *Research Design: Qualitative, Quantitative, and Mixed Methods Approaches*. 2 ed. Thousand Oaks, CA: Sage.

Dunn, K. 2005. Interviewing, in *Qualitative Research Methods in Human Geography*, edited by I. Hay, 79-105. Oxford: Oxford University Press.

Elden, S. 2007. Governmentality, calculation, territory. *Environment and Planning D: Society and Space* 25(3): 562-580.

Elwood, S.A. and D.G. Martin 2000. 'Placing' interviews: Location and scales of power in qualitative research. *Professional Geographer* 52 (4): 649-657.

Forest, B. 2001. Mapping democracy: Racial identity and the quandary of political representation. *Annals of the Association of American Geographers* 91(1): 143-166.

Gill, N. 2009. Governmental mobility: the power effects of the movement of detained asylum seekers around Britain's detention estate. *Political Geography* 28(3): 186-196.

Gilmore, R.W. 2007. *Golden Gulag: Prisons, Surplus, Crisis, and Opposition in Globalizing California*. Berkeley: University of California Press.

Gilmore, R.W 2008. Forgotten places and the seeds of grassroots planning, in *Engaging Contradictions: Theory, Politics, and Methods of Activist Scholarship*, edited by C.R. Hale. Berkeley, CA: University of California Press, 141-162.

Gilmore, R.W. and Gilmore, C. 2008. Restating the obvious, in *Indefensible Space: The Architecture of the National Insecurity State*, edited by M. Sorkin. New York: Routledge, 141-162.

Groves, R.M. 2010. *Prepared Statement of Robert M. Groves, Director US Census Bureau: 2010 Census: Enumerating People Living in Group Quarters*. Washington, DC: US Department of Commerce, Bureau of the Census.

Gupta, A. 1995. Blurred boundaries: the discourse of corruption, the culture of politics, and the imagined state. *American Ethnologist* 22(2): 375-402.

Hannah, M. 2000. *Governmentality and the Mastery of Territory in Nineteenth Century America*. Cambridge: Cambridge University Press.

Hannah, M. 2001. Sampling and the politics of representation in US Census 2000. *Environment and Planning D: Society and Space* 19(5): 515-534.

Hannah, M. 2009. Calculable territory and the West German census boycott movements of the 1980s. *Political Geography* 28(1): 66-75.

Ho, D.E. 2011. Captive constituents: prison-based gerrymandering and the current redistricting cycle. *Stanford Law & Policy Review* 22(2): 355-394.

Hodder, I. 2000. The Interpretation of Documents and Material Culture, in *Handbook of Qualitative Research*, edited by N.K. Denzin and Y.S. Lincoln. Thousand Oaks, CA: Sage, 703-716.

Janesick, V.J. 1999. A journal about journal writing as a qualitative research technique: History, issues, and reflections. *Qualitative Inquiry* 5 (4): 505-524.

Janesick, V.J. 2000. The choreography of qualitative research design: Minuets, improvisations, and crystallization, in *Handbook of Qualitative Research*, edited by N.K. Denzin and Y.S. Lincoln. Thousand Oaks, CA: Sage, 379-400.

Johnston, R. and C. Pattie 2003. Representative democracy and electoral geography, in *A Companion to Political Geography*, edited by J. Agnew, K. Mitchell and G. Toal. Malden, MA: Blackwell, 337-356.

Kelly, J.P. (forthcoming). The strategic use of prisons in partisan gerrymandering. *Legislative Studies Quarterly*.

Levingston, K.D. and C. Muller 2006. 'Home' in 2010: A report on the feasibility of enumerating people in prison at their home addresses in the next census. New York: Brennan Center for Justice at NYU School of Law.

Libal, B., L. Martin and N. Porter (2012). A prison is not a home: notes from the campaign to end immigrant family detention, in *Beyond Walls and Cages: Prisons, Borders, and Global Crisis*, edited by J. Loyd, M. Mitchelson and A. Burridge. Athens, GA: UGA Press.

Lotke, E. and P. Wagner 2005. Prisoners of the census: Electoral and financial consequences of counting prisoners where they go, not where they come from. *Pace Law Review* 24(2): 587-607.

Marantz, A. 2006. The five-fifths clause: how we count, and use, our prisoners. The Washington Post Company [Online 6 November]. Available at http://www.slate.com/id/2152994 [accessed 2 February 2007].

Martin, L. and M.L. Mitchelson 2009. Geographies of detention and imprisonment: interrogating spatial practices of confinement, discipline, law, and state power. *Geography Compass* 3(1): 459-477.

McKittrick, K. 2012. On plantations, prisons, and a black sense of place. *Social & Cultural Geography* 12(8): 947-963.

Mitchelson, M.L. 2012. The urban geography of prisons: mapping the city's 'other' gated community. *Urban Geography* 33(1): 147-157.

Moran, D., L. Piacentini and J. Pallot, J. 2012. Disciplined mobility and carceral geography: Prisoner transport in Russia. *Transactions of the Institute of British Geographers* 37(3): 446-460.

Morrill, R.L. 1981. *Political Redistricting and Geographic Theory*. Washington, DC: Association of American Geographers.

National Research Council. 2006. *Once, Only Once, and in the Right Place: Residence Rules in the Decennial Census*, edited by D.L. Cork and P.R. Voss. Washington, DC: The National Academies Press.

New York Times Editorial Board. 2006. Prison-Based Gerrymandering. *The New York Times*, 20 May:A12.

Nobles, M. 2000. *Shades of Citizenship*. Stanford, CA: Stanford University Press.

Prison Policy Initiative. 2011. [Online: Publications Archive]. Available at http://www.prisonpolicy.org [accessed: 12 December 2011].

Prisoners of the Census. 2012. *The Solutions.* [Online]. Available at http://www.prisonersofthecensus.org/solutions.html.

Richards, S.C. and J.I. Ross. 2003. A convict perspective on the classification of prisoners. *Criminology & Public Policy* 2 (2): 243-252.

Roberts, S. 2008. Census Bureau's counting of prisoners benefits some rural voting districts. *The New York Times*, 24 October A12.

Roberts, S. 2010. New option for the States on inmates in the census. *The New York Times*, 11 February A18.

Rose, N. 1999. *Powers of Freedom*. New York: Cambridge University Press.

Rose-Redwood, R.S. 2006. Governmentality, geography, and the geo-coded world. *Progress in Human Geography* 30(4): 469-486.

Rose-Redwood, R.S 2008. Indexing the great ledger of the community: Urban house numbering, city directories, and the production of spatial legibility. *Journal of Historical Geography* 34: 286-310.

Scott, J.C. 1998. *Seeing Like a State*. New Haven, CT: Yale University Press.

Soja, E.W. 1980. The socio-spatial dialectic. *Annals of the Association of American Geographers* 70(2): 207-225.

Starkweather, S. 2009. Governmentality, territory and the U.S. census: The 2004 Overseas Enumeration Test. *Political Geography* 28: 239-247.

Stinebrickner-Kauffman, T. 2004. Counting matters: Prison inmates, population bases, and 'One Person, One Vote'. *Virginia Journal of Social Policy & the Law* 11: 229-253.

Tonry, M. 1995. *Malign Neglect: Race, Crime, and Punishment in America.* Oxford: Oxford University Press.

U.S. Census Bureau. 2006. U.S. Census Bureau Report: Tabulating Prisoners at Their 'Permanent Home of Record' Address. Washington, DC.

Wacquant, L. 2001. Deadly symbiosis: When ghetto and prison meet and mesh. *Punishment and Society* 3(1): 95-134.

Wagner, P. 2010. *Testimony of Peter Wagner, Executive Director, Prison Policy Initiative, before the Information Policy, Census, and National Archives Subcommittee of the Oversight and Government Reform Committee, February 22, 2010*. Northampton, MA: Prison Policy Initiative.

Wagner, P., E. Lotke and A. Beveridge 2006. *Why the Census Bureau can and must start collecting the home addresses of incarcerated people*. Northampton, MA: Prison Policy Initiative.

Webster, G.R. 2004. Representation, geographic districting, and social justice. *Journal of Geography* 103(3): 111-126.

Chapter 7

Landscapes of Toxic Exclusion: Inmate Labour and Electronics Recycling in the United States

Kelsey Nowakowski

In United States popular culture, prison workers are commonly depicted as contributors to public works programmes, generally in the realm of license plate making and road waste removal. Today, however, thousands of inmates in the US find themselves employed by a private entity, as 31 per cent of all state and federal correctional facilities offer private sector jobs to prisoners (Stephan 2008). Outlawed due to union protests in the 1930s and 1940s, the current revival of private sector use of prison labour came in 1979, when the US Congress lifted its ban on the interstate transportation of prisoner-made goods. Since then, prison labour has become much more integrated into the economy than one might originally suspect by way of private ventures, such as furniture manufacturing, call desk support, and electronics recycling.

The privatization of inmate labour, as well as that of prisons as a whole, leads many to believe that the prison system is more concerned with lucrative profits than it is with rehabilitating convicts. States find private facilities attractive, presumably, because they help reduce the economic costs of incarceration. In addition to the privatization of carceral institutions, states have also turned increasingly toward inmate labour to subsidize incarceration costs. In both public and private prisons alike, inmates are employed in custodial work, which includes cooking and cleaning responsibilities, as well as manufacturing jobs for private entities. The private companies that hire inmates at both private and state institutions, however, are not subject to the same laws that typical businesses are. In fact, legally, they fly below the radar, as they are not required to have surprise inspections conducted by the Occupational Safety and Health Administration (OSHA). As such, a great deal of exploitation can – and does – go unnoticed in these facilities. This chapter will argue that not only do inmate workers receive unfair compensation for their labour, but that some work programmes also increase their likelihood of injury and of developing a health problem, as safety standards are often disregarded.

Beginning with a brief overview of the historical-geographical development of prison labour, the first section aims to show how the labour of inmates has been used to fit the economic and political conditions of the time and region. The chapter then moves toward a case study that shows how the goods inmates

produce are extremely mobile despite the remoteness and restrictedness of carceral spaces. Although inmates represent a socially excluded population, the goods they produce circulate through the market and in effect integrate their labour into the wider economy. The integration of their labour into the economy is premised on exclusion, because labour laws do not protect prisoners and they live in spaces hidden from the public eye. Inmates are confined to a given carceral space, but the mobility of the goods they produce connects them to a wider economic circle.

As states struggle to handle the ever-growing waste stream, obsolete electronics often end-up in prisons where they are dismantled and recycled. During the electronics recycling process, prisoners and staff at Atwater Federal Prison in Merced, CA, have been routinely exposed to toxic heavy metals. It is difficult to enforce safety laws in prisons and to recognize when abuses are taking place, because carceral spaces are notably obscured from the public eye. The social and political exclusion of inmates encourages companies to employ prisoners in the most hazardous steps of the commodity chain as they save money by not having to follow strict regulations. For this reason, I argue that inmates are integrated into the economy on the basis of their exclusion from society and the laws that regulate work. Prisons, then, become 'landscapes of defence', because inmate labour defends the rest of society from the harmful by-products of the commodity chain, as will be discussed below.

The Historical-Geographical Development of Inmate Labour

Inmate labour has adapted over time to serve the respective economic system and political condition. In the United States, inmate labour has been in use since the late eighteenth century, yet it was not heavily relied upon until the abolition of slavery in 1865. Since emancipation, prison labour has been utilized in many ways and by various entities, such as plantation owners, state and federal governments, and private companies. Today, the fastest growing use of inmate labour is in the private sector (Rice 2006), but this has not always been the case. Up until the late 1920s, private companies were free to employ inmates and transport inmate-made goods across state borders without restriction. By 1935, however, labour union opposition prompted the government to pass laws limiting the use of inmate labour and the interstate trading of prison-made goods. These laws were repealed by the Reagan administration; inmate labour was once again accessible to the private sector as conservatives believed it was a way to promote economic efficiency. As such, the United States government has oscillated between promoting and limiting the use of inmate labour in the private sector. Although restricted at times due to changes in the political economy, inmate labour has always been appropriated to perform cost-effective work.

After slavery ended in 1865, the political economic system of each region determined what type of work prisoners performed and how they were housed. In the South, where an agricultural economy predominated, private contractors used

inmate labour to uphold as much as possible the plantation economy that thrived under slavery. Housing prisoners on the plantation under the convict lease system, even while having to pay the state for their labour, allowed former slave owners to easily run their farming operations and continue making high profits. In contrast, in the North, where a manufacturing economy developed, contractors were only interested in managing prison workshops so they could manufacture goods and lower costs. The goal in each region was the same: to amass high profits; however, in the North manufacturers were able to defray production costs even more by not providing housing or food for inmates.

The 'state use system', as the use of inmate labour for public works came to be called, was adapted to meet the labour needs of the South in two ways (Ryan and Ward 1989). First, convict labour was used almost exclusively outside prison grounds on public works such as the construction of bridges and roads. This practice transformed the lease system into the notorious chain gang road crews characteristic of the landscapes of the South. Second, inmates at this time began to work on penal farms in states such as Florida and Louisiana under the supervision of overseers that showed ruthless disregard for human rights, making these practices very akin to slavery. Although these inhumane conditions were protested by some organizations, they continued in the South until the 1950s.

In 1934, under Roosevelt's administration, a corporation called Federal Prison Industries, Inc. (FPI) was created as an arm of the Federal Bureau of Prisons (Atkinson 2002). Known most commonly today as UNICOR – an arbitrary acronym – FPI contracts prison labour to both the public and private sector to make goods for use by the state and private consumers, such as office furniture and electronics. According to Hallett (2006), FPI has a legal monopoly because mandatory contracting rules make it illegal for government agencies to solicit bids from outside contractors for goods and services already provided by FPI, which leads to several advantages.

Union protests of the 1930s led to fewer inmates working in manufacturing. From the 1940s to 1970s, most prisoners performed custodial jobs. At this time, there was an increase in inmate education and rehabilitation therapy, as liberal activists successfully instituted welfare programmes in prisons (Bair 2008). For this reason, this period came to be known as the 'Treatment Era', a time when rehabilitation programmes were stressed over work programmes (Bair 2008: 121). Yet, many conservatives still advocated for the use of inmate labour, because they saw the practice as being beneficial to both the inmate's re-education and the state's finances. To create more jobs for inmates, Congress passed the Justice System Improvement Act in 1979 and revived the use of prison labour in the private sector, albeit setting up many restrictions on the practice. Under this law the Private Sector/ Prison Industries Enhancement Certification Program (PIE) was launched, allowing for private industry to 'make bids for inmate labour-power, creating the conditions for partnerships between the state and private enterprises' (Bair 2007: 88).

Not only did the Federal Bureau of Prisons justify opening up to the private sector by boasting of the increased work opportunities it provided inmates, but it also advertised its practices as a way to decrease the cost of an individual's incarceration. Because working inmates contributed to their own room and board, family support, victim compensation, and state taxes, it was argued that society benefited at large from the reduction of incarceration costs (PIECP 2004). Outwardly, companies and the Bureau of Prisons (BOP) justified employing inmate workers based on: 1) the belief that only private businesses can create enough jobs to occupy prisoners' time, 2) research supporting the idea that being employed will give prisoners valuable skills so they are able to obtain jobs when released, and 3) the idea that prisons are expensive to run so private manufacturers can contribute to making them more affordable by paying the state to operate on prison grounds. This reopening to the private sector, however, has not been void of resistance. Although the Bureau of Prisons maintains that it consults local labour unions before putting inmates to work, there have still been cases of protest. For example, in 2003 former workers from Dell's Computer Recycling Team protested the loss of their jobs to inmate labour. Dressed in mock prisoner attire they chanted: 'I lost my job. I robbed a store. Went to jail. I got my job back' (CNN 2003). More recently, groups of Louisiana and Alabama fisherman have protested British Petroleum's use of inmate labour to clean up the 2010 Gulf Oil Spill, citing the fact that the company would rather hire cheap convict labourers than employ out-of-work men and women. Clearly, the revitalization of inmate labour in the private sector has its critics. Yet, some communities, especially in rural areas, welcome new prisons, because of the potential economic opportunity they offer. Anne Bonds (2006) studies this trend in the American Northwest and questions why prisons represent a 'recession-proof strategy for rural community development' amidst a political-economic and cultural climate of rural population decline and neoliberal disinvestment. In addition to spurring job growth, prisons can increase local tax revenues, because inmates are counted as residents in the communities where they are incarcerated.

The move to privatize both prisons and prison labour must also be put within a greater economic and political context and cannot simply be explained by the reasons discussed above. The harsher sentencing policies of the War on Drugs era led to a massive increase in incarceration rates. This prompted the Bureau of Justice to order 60 per cent of states to reduce prison overcrowding. For this reason, the 1980s represent a time of massive prison building programmes, with inmates in some cases helping to provide the labour for these constructions (Ryan and Ward 1989). This large increase in the rate of incarceration shows that the number of people behind bars increased at a far more rapid rate in proportion to the overall population of the US, demonstrating why there was an increasing burden on state budgets to run and build more prisons.

The Reagan, Bush Sr., and Clinton administrations supported the privatization of prisons and prison labour. In fact, private incarceration began with a Reagan-sponsored experiment to house Immigration and Naturalization Service (INS)

detainees at private detention centres in Texas (Parenti 2001). According to David Lapido (2001: 118-19), privately built and run prisons have been operating under state and federal government contracts since the early years of the Reagan administration and have been supported through the propagation of the belief that 'free-marketeers urge cutting taxes and public expenditure, a more managerialist, businesslike economy'. Private prisons have precisely this effect; they provide revenue from the running of the prison, which would otherwise be paid for by taxpayers. Prisoners being required to pay for their incarceration marks a stark shift from the rehabilitative model used from the 1940s to the1970s to a for-profit model in use today. Prisons, thus, are now commonly run as private enterprises, adhering to the logics of cost-cutting and profit motive. Possessing a deeper understanding of the economic policies that led to the revitalization of private sector involvement in prisons allows one to understand how inmate labour has become increasingly integrated into the economy. The austere economic policies encouraged by the Reagan administration can be viewed as the means by which private companies were enabled to become reliant on inmates for cheap labour.

While this chapter is US centric, it is important to explore the differences in how inmate labour is appropriated in other countries. In China, inmates reportedly work for up to 14 hours a day with no pay in the 're-education through labour' system. Former inmates claim these labour camps are highly militarized as the Chinese government believes prisoners can solve their personal problems through strictly organized physical labour (Luard 2005). American trade unions criticize these Chinese penal work camps for being both inhumane and for producing unfairly competitive goods, but they often do not realize similar working conditions exist in US prisons. In Norway, prison models function much differently than most countries, as inmates are still granted the right to vote and to participate in political debates. Most Norwegian prisoners finish their sentences in what is termed an 'open prison', meaning they have free mobility around the prison (Fouche 2009). In this system, prisoners prepare for re-entry with more liberty and are given work assignments, such as farming and carpentry, skills that prepare them for employment upon release. US prisoners cannot vote and it is often debatable how marketable the skills they learn from their work assignments are, which is particularly the case for inmates who are employed in UNICOR's electronics recycling facilities.

A Case Study: Recycling Electronic Waste in Prisons

Electronic waste – computers, monitors, copy machines, cell phones, etc., – is now the fastest growing sector of the municipal waste stream, rising at a rate of five per cent a year (Dayaneni and Shuman 2007). Until recently, most electronic waste went directly into landfills or was shipped to Asia and Africa, especially China, India, and Nigeria, for both recycling and disposal. During the 1980s, industrialized countries tightened waste legislation, thereby increasing the

cost of domestic disposal and creating a financial incentive to export waste for recycling (Lepawsky and McNabb 2009: 3). Many devices are not designed to be easily disassembled and refurbished, so their recycling process can be both time-consuming and costly. Since valuable parts of old electronics are difficult to recover because of poor design, cheap labour, often found overseas, is required to perform manual disassembly (Davis and Smith 2003). For this reason, a great deal of electronic waste, also known as e-waste, is handled and processed by the most socioeconomically marginalized groups of people. But not all e-waste is sent overseas as a number of prisoners have found themselves in the recycling line of business.

In 2008, there were 21,836 inmates (17 per cent of all medically able inmates) working in all FPI programmes, which include furniture and electronics manufacturing, textile and clothing production, and electronics recycling, at 107 factories located at 76 prisons around the country (FPI 2009). According to UNICOR's website, UNICOR is a vocational training programme designed to enhance the education and work skills of inmates, as well as decrease the cost of an inmate's stay in prison by deducting a portion of his or her earnings and applying it to incarceration fees. UNICOR works to pursue new business strategies in hopes of repatriating jobs from overseas sweatshops, thereby creating partnerships in which both UNICOR and the domestic private sector simultaneously benefit (Jackson et al. 2006: 20-21). UNICOR's website also states that former inmates who work in their prison industries or who complete vocational apprenticeship programmes were 24 per cent less likely to recidivate than non-programme participants and 14 per cent more likely to be gainfully employed. This data, however, fails to address the differences between working in the prison industries and the apprenticeship programmes, as well as the limited number of people who have the opportunity to work as apprentices and be properly trained.

With prison recycling programmes, UNICOR can successfully compete with developing countries that accept e-waste. Opening its first electronics recycling factory at a federal prison in Marianna, Florida in 1994, UNICOR has since spread its operation to seven other federal prisons across the country. At this facility, female prisoners unload truckloads of computers each day and break them down for parts, such as processors or cathode ray tubes (CRTs), which can be reused or sold. UNICOR's objective is to derive the highest and best use for the items donated for recycling. Inmates are paid from 23 cents to $1.15 per hour, which is what they earn after 50 per cent of their wages are contributed to: child support, alimony, court fines, and victim restitution as ordered by a court of law. With this little pay going to inmates, UNICOR is able to compete with private recycling companies while keeping prisoners busy.

Set up in 1934 to provide goods and services exclusively to the federal government, UNICOR has long since transformed into a profit seeking company that regularly does business with the private sector. UNICOR generates revenue from the sale of more than 175 diverse products and services that it sells to the federal government and many other private businesses and organizations (FPI 2009).

When UNICOR solely conducted business with the government prior to the 1979 Justice System Improvement Act, it may have been accurate to consider inmate labour separate from the competitive economy, as private companies were then not bidding against prison labour for contracts with other companies. Since the 1980s, as discussed above, inmate labour has become increasingly integrated into the private sector. UNICOR officially does business with state and city governments, federal agencies, private industries, and non-profit organizations. Inmate labour, thus, is integrated into many different markets.

While some prisoners appreciate the work they conduct for FPI, others do not. For the inmates who are pleased about being employed, the most common reasons they cite are that the work gives them purpose, keeps them preoccupied, and puts a little extra money on their books. Some prisons require inmates to do custodial chores even when these jobs are not sought out, while most industrial and private sector jobs are in demand, because they pay slightly more and are often times more interesting work. The money inmates earn, maybe $30 dollars in a month, can be used in a handful of ways, such as for extra food, hygiene products, or for court fines. For inmates planning their release, they may save their earnings to assist with expenses when they leave the prison. Yet, there are also US prisons where inmates are less enthusiastic about the jobs they perform. In some carceral institutions, inmates are not compensated for their work, which was a major factor that led to the 2010 protests in Georgia state prisons. Inmates in at least seven Georgia prisons refused to perform chores, do industrial jobs, or shop at prison commissaries (Wheaton 2010). Through their non-violent strike, inmates demanded that they be paid for their work and have more educational opportunities, as well as improved living conditions. In other instances, prisoners have been routinely exposed to toxicities in their work environment, but have been unable to voice their complaints. As will be evidenced below, it took a Bureau of Prisons employee to champion for the rights of inmates working in the electronics recycling industry.

The Integration of Inmate Workers in the Economy

Circulation is commonly conceived of as a closed system, but in economics it is connoted as a movement of value ending in capital accumulation (Swyngedouw 2006). Representing a process of change and transformation, it is during circulation that resources are altered in order to produce useful commodities. Also conceptualized as a spatial flow, the notion of circulation denotes a practice whereby goods and the resources used to create them travel on and are transformed within commodity chains. As commodities circulate through the production chain, the labour of workers who are creating these goods becomes integrated into the wider economy. For this reason, even isolated spaces can become part of a larger economic sphere, or global chain of production.

Extensive material flows exist that embed prisoners in the economy by way of the electronics recycling industry. Prisoners become part of the larger commodity chain during the disposal and recycling process, as their role in the division of labour is to demanufacture discarded electronics. Reprocessed electronic parts are then sold on the market to be used again in the production of new electronic commodities. This entire sequence of circulation is carried out for the sake of earning profit, or as Swyngedouw (2006) would explain, to accumulate capital. In this manner, inmates play a major role in the commodity chain and the accumulation process. As such, the goods they produce have a much greater mobility than one would originally suspect.

Although most assume the spatial segregation of inmate labour, prisoners that work for UNICOR are deeply integrated into the economy. UNICOR provides recycling services to federal, state and local governments, private-sector businesses and not-for-profit agencies and relies on recyclers and reprocessors to collect and transport the electronic goods and parts (Skillings 2002). Moreover, UNICOR is ranked among the nation's top processors of used electronics (Vanderpool 2006a), as electronics recycling is its fastest-growing business with some 900 inmate workers processing nearly 44 million pounds of electronic equipment per year, or 50,000 pounds per prisoner per year (Grossman 2005). In September 2002, for example, the firm had 833 inmates at seven federal facilities engaged in electronics recycling with net sales of 3.4 million dollars. One year later, the net sales for recycling had more than doubled to just over eight million dollars (Reid 2004). In 2008, with 876 inmates employed by its recycling programme, UNICOR netted 10.5 million dollars in sales, processing more than 37 million pounds of obsolete and excess electronics from both the public and private sectors (FPI 2009). UNICOR, as made apparent by the above information, is a growing business connected to more profit-seeking entities than just the US government. Other electronics recycling companies complain that they cannot compete with UNICOR, because, unlike UNICOR, they must charge to recycle discarded materials in order to pay their workers a fair wage.

One major way UNICOR makes itself financially attractive to so many corporations and organizations is by regionalizing its factories such that at least one is accessible to each part of the country. Of the top 10 reasons to buy products from UNICOR according to its website, number six is the fact that it has nationwide coverage: 'Accessible, competitive and right next door! Our nationwide network of manufacturing locations and support optimizes order tracking, communication, transportation and savings' (www.unicor.gov). Eight processing centres and six collection sites are located throughout the country, thereby giving adequate access to potential donors of electronic scrap and customers to buy it once is refurbished. According to UNICOR, nationwide locations allow for exceptional value and competitive pricing, and UNICOR competes with private recycling entities in each and every region of the country. By accepting donations from public organizations, private companies, and government agencies UNICOR receives a steady stream of outdated electronic equipment. So, it is clear that the electronic goods that inmates

recycle have an extensive mobility as they make their way to the prison factory, as well as when they are sold to other entities once refurbished.

Prisons as Landscapes of Defence

Dangerous segments of the production and waste disposal process are often confined to specified areas in order to defend the rest of society from the harmful by-products of the commodity chain. Conceptualizing prisons and other workplaces as 'landscapes of defence' shows how certain spaces are exploited in order to keep more affluent spaces safe. By this, I mean risks are contained in hidden landscapes like prisons to ensure the well-being of privileged members of society. Landscapes, thus, have the ability to ensure security for some while creating insecurity for others. In *Landscapes of Defence*, Gold and Revill (2000: 3) define defence as a term conceptually related to security, minimization of risk, and the preservation of power, and landscape as a topographic interpretation of a designated portion of space. They coined the term landscapes of defence to show how certain spaces are defended to protect the people who inhabit them, and how some spaces are designed to minimize risk in other areas. That is, there is an attempt to 'fix' risk in a designated area, so that a more advantaged area does not have to deal with harmful by-products. Hidden locales, like prisons, make an ideal landscape of defence, because the political and social climate allows for exploitation.

Prisons with recycling operations function as landscapes of defence and integrate inmates into the economy in a questionable manner. Atwater Federal Prison, a maximum-security institution located outside the city Merced in California's Central Valley, provides a telling example of how inmates are at once integrated into the economy and divided from society. Operating an electronics recycling plant at Atwater since 2002, UNICOR's factory has had numerous safety problems and closures. Because this prison recycling factory acts as a landscape of defence, inmates here are subjected to the most hazardous circulations of the commodity chain, as labour laws do not protect them.

When e-waste arrives at a UNICOR prison factory, it is searched at a warehouse facility located on the outer perimeter of the main prison. Approximately 40 low risk inmates work at this warehouse where they restack computer monitors and televisions before sending them to the main processing facility. Higher risk inmates, who often serve longer sentences, do the bulk of recycling in the facility, so UNICOR has lower worker turnover. Electronics are then put into one of two tracks. Old devices will be cleaned up for resale, meaning their hard drives will be wiped, while other non-functioning devices will be mined for materials which include glass, plastics, and copper wiring (Skillings 2002). According to Jackson et al. (2006: 33), electronic materials in prisons are handled up to three times more than by properly regulated recycling facilities.

Furthermore, UNICOR's recycling programme is labour intensive so equipment expenses are kept low. Because UNICOR employs a captive labour force, it does not have to offer competitive wages and can thus, make higher profits that its competitors. Yet, not having the proper machinery to disassemble electronic parts exposes inmates directly to toxic materials. Often times, inmates were denied access to the proper type of tip to unscrew computer monitors, so they would be forced to break the leaded glass with hammers (Jackson et al. 2006: 9). By busting open the screen on televisions and monitors, the CRT also breaks and releases lead and cadmium dust into the air (Sample 2009: 4). In addition, using hammers increases the risk of injury from shattered glass and plastic shards. Prisoners would break CRTs inside television and computer monitors without wearing masks, gloves, or other protective gear, which in turn exposed them to heavy metal dust. While responsible private recycling companies use a closed-system mechanical crushing machine, UNICOR dismantles CRTs by having eight-12 inmates smash them with a hammer in a caged area blocked off with strips of plastic sheeting.

Inmates' exposure to such hazards, however, was not only limited to the times they were working. In order to limit inmate mobility during work hours, which would be cause for added security, UNICOR workers ate their meals in contaminated areas in the factory (Stade 2005). The coveralls and boots provided for inmates to wear while working were also worn during break time and when eating meals, so prisoners often tracked toxins to other places. In legitimate recycling operations, the Occupational Safety and Health Administration (OSHA) avoids such hazardous practices by ensuring workers change their contaminated clothing before leaving their place of work. Yet, neither prison staff nor inmates were informed of such health risks and no training on the handling of contaminants was provided to them (Stade 2005). For all of these reasons, Leroy Smith, a Federal Bureau of Prisons employee for over a decade, filed a complaint with OSHA claiming that inmates at Atwater Federal Prisons computer recycling programme had been deliberately exposed to toxicities.

Smith claims that inmates used equipment not up to OSHA's standards and worked in conditions that would not be permitted in the private sector. Smith told Grossman (2006: 249) in a 2005 interview that there was not sufficient training, guidance, or proper procedures put in place, which resulted in a mass of injuries. After experiencing symptoms of hazardous material exposure, including anxiety, headaches, fatigue, memory lapses, skin lesions, and circulatory and respiratory problems, Smith paid out of his own department's budget to have air quality testing done because the Bureau of Prisons (BOP) refused to pay for the inspections (Sample 2009: 2). The results of these tests found cadmium and lead levels in excess of OSHA standards, resulting in the first shutdown of UNICOR's Atwater facility on July 1, 2002 (Jackson et al. 2006: 12). In addition, wipe samples found lead, cadmium, and beryllium on inmates' skin and work surfaces. In December 2004, after being ignored by superiors at Atwater, Smith filed a complaint with OSHA. Just five months later, OSHA granted Atwater a clean bill of health despite UNICOR failing to implement meaningful safety improvements

(Vanderpool 2006b). Frustrated, Smith then contacted the US Office of the Special Council (OSC), an independent federal investigative agency that is responsible for protecting federal employees and reviewing whistleblowers' complaints. After inspecting Atwater, Special Counsel Scott Bloch confirmed Smith's complaints, which contrasted both OSHA and BOP reports. In an April 2006 report, Special Counsel Scott Bloch (in Vanderpool 2006a) wrote that prison managers recklessly, and apparently in some cases knowingly, exposed inmates and staff to unsafe levels of hazardous materials over several years.

Since the OSC's release of its findings, the federal government has undergone further investigation of Atwater's recycling operation. According to a preliminary report issued by the Justice Department's Office of the Inspector General (OIG) and an evaluation issued by the Federal Occupational Health Service in October 2008, FPI has jeopardized the lives and safety of many prisoners and staff (Sample 2009: 1). The report also called for many issues to be addressed, including ensuring there is full compliance with all applicable OSHA standards, performance of a detailed job hazard analysis prior to beginning any new operations or when making changes to existing ones, and making sure inmates have a way to voice their concerns about how to improve their workspace. Even with the given safety recommendations, prisons with these electronics recycling factories continue to function as landscapes of defence, which sets up a particularly significant tension. Dikeç (2005) addresses this dilemma by relating a fictional homeless man's story about what the best method is to solve a prison food shortage. After discussing this issue, the man then presents a more thought provoking question: is the prison system and how it functions in a broader context the actual problem? Likewise, is it best to improve the working conditions for these workers or does this in some way condone the wrongdoings that took place? Were those in charge held accountable and who is monitoring if improvements have been made? Perhaps the way the entire prison system functions is to blame for the mistreatment of inmates, as the proper measures have never been put in place to make it a fair institution.

Sites of Political Exclusion and Social Marginalization

UNICOR's prison recycling factories are not open workplaces and the safety violations UNICOR has committed against inmate workers have been largely obscured from the public. Since inmates are guaranteed very few workers' rights, it is easy for prison industries to take advantage of their legal marginalization. According to UNICOR's recycling website, it is 'the responsibility of each Federal Agency to establish and maintain an effective and comprehensive occupational safety and health programme'. Inmates, then, fall outside the realm of legal protection, as OSHA is unable to conduct surprise inspections in prison factories. Can safety laws be sufficiently enforced if OSHA does not officially oversee UNICOR? Who ensures that UNICOR will adhere to standards? OSHA, the EPA, state, and federal agencies allow inspections, but because these entities must give

advance notice to the prison before coming, UNICOR facilities have time to clean up their dirty operations. In fact, sworn statements from prisoners and guards detail 'such things as being ordered by prison staff to work more slowly, to clean more thoroughly than usual, to change air filters, and even to wear safety equipment provided for the first time prior to the inspection' (Shuman 2007: 6). So, unlike private companies, UNICOR has time to tidy its operation so as to avoid paying fines or being shut down. Unfortunately, regulatory agencies cannot document or test real work conditions without the right to conduct surprise inspections of UNICOR's recycling programme. Regulatory agencies like OSHA, therefore, fail to adequately protect inmate labourers from the abuses of their employer. Because inmates do not have an established channel to voice complaints through, there have been no strikes organized by them within the prison. As is evidenced in this case study, it took a BOP worker's protests to gain the attention of regulating officials.

According to a US Chamber of Commerce (2005) statement on FPI's contracting practices, UNICOR is exempt from federal and state income taxes, gross receipt taxes, excise tax, and state and local sales taxes on purchases. UNICOR also receives taxpayer support for a number of subsidies, including land and buildings, making its operation expenses very low (Dayaneni and Shuman 2007). Another way UNICOR saves money is by not providing workers with benefits like social security, unemployment compensation, or medical insurance. These findings serve to show inmates are outside the protection of the law, because they are not considered a part of the normal labour realm. Inmate workers are not permitted to file a compensation claim for injuries until within 45 days of their release, and although they have a right to legal representation, they cannot require anyone to testify and must pay all the expenses of presenting evidence, including the costs of travelling to Washington DC for hearings (Shuman 2007: 9). Because inmates have no viable avenue through which to voice complaints, UNICOR is able exploit its prison labourers without fear of legal repercussions. Prisoners, perceived as a risk to society, become a particularly vulnerable working population.

UNICOR fails to give workers access to information and safety advice about hazardous materials, as well as denies workers the right to protest without fear of being fired. Private recycling companies must follow all federal, state, and local safety standards and regulations, because they can receive surprise inspections from OSHA at any time. Unfortunately, prisons are not subjected to the same laws. When prisoners question UNICOR's health and safety practices, they fear being fired, punished, or moved to another prison. Although their labour is just as much a part of the electronics recycling industry as is that of workers for a private company, inmates are neither protected from the hazardous risks of the electronics waste stream, nor from the practices of an unregulated industry.

Using prisoners to recycle hazardous waste represents a serious case of social discrimination, one that asserts that the health and safety of prisoners is not as important as the well-being of free citizens. Prisoners are a group of people assumed

to be dangerous to society because of the crimes they have committed. The manner in which inmate labour is integrated into the economy, thus, raises many questions of environmental justice and waste politics. Environmental justice is grounded in the principle that all communities have a right to a healthy environment, despite their social standing (Jackson et al. 2006: 26). Inmates, however, have been denied a healthy workspace because of their legal status. In effect, they have been exposed to the most harmful circulations of the e-waste cycle. There is little objection when risks, in the form of waste, are handled in areas where marginalized people, like inmates, reside. Society's risks, in this case electronic waste, are increasingly externalized onto the rising population of captive prison labour.

The role a person plays in the division of labour is very telling of his or her social condition. As Lepawsky and McNabb (2009: 23) argue, 'it is possible to glimpse how the labour of e-waste processing is constitutive of broader social geographies of belonging and marginalization'. Sending e-waste to prisons reflects a 'willingness to displace the burden of toxics onto the people least empowered to advocate for their own safety' (Shuman 2007: 1). Enforcing health and safety regulations is most difficult in prisons, so inmates are forced to remain in their marginalized position. Inmates, unlike free workers, must labour in dangerous conditions because their political exclusion from society makes their health and safety less of a concern to regulating agencies.

Conclusion

Conceptualizing Atwater Federal Prison as a landscape of defence demonstrates how certain landscapes are relied upon as spatial fixes. Because electronics contain a great amount of toxic components, they must be disposed of safely. Handling hazardous chemicals is not something most people are willing to do, as toxicities can cause serious and permanent damage to human health. Prisoners, because of their social marginalization and spatial isolation, are employed in disassembling electronics without much regard to the effects of hazardous chemicals on their well-being. Prisons, thus, not only provide security for the rest of society by keeping (presumably) dangerous criminals behind bars, but they also do so by handling the most hazardous circulations of the electronics waste stream.

The labour of inmates defends the rest of society from the risks of chemical waste, and this exposed labour will play an increasingly important role in the peculiar division of labour as the consumption of obsolete electronics grows. The most toxic part of post-production takes place in prisons because of a general lack of objection to these practices being carried out by inmates *seen as criminals*. Because inmates are excluded from equal workers' rights and fair wages, businesses save money in employing them by cutting the costs that would secure safe labour practices in addition to paying low wages. Contractors that employ inmates do so because they save money by not having to pay for proper equipment and training, as well as medical compensation. The integration of inmate labour

into the external economy, then, is premised on their social exclusion. Prison labourers are attractive to businesses precisely, because they can be abused without legal repercussion.

This example of UNICOR's recycling programmes shows how the labour of inmates is incorporated into many markets, because of the mobility of the goods they produce. As noted in the historical section, this current privatization trend is deeply founded in 1980s Reagan-era politics that stressed fiscal austerity. Although the general notion of inmates working can be viewed as beneficial, the type of work some inmates carry out is extremely harmful to their health and should not be overlooked. For this reason and despite prisons generally being located in remote areas, it is necessary for the events that take place in carceral spaces to be more transparent, as well as more accountable to standard safety laws. Only then will inmate labourers be able to overcome their political and social exclusion. The US prison system could benefit by studying countries with rehabilitative prison models, especially that of Norway, which functions to successfully rehabilitate and transition inmates back to society by providing them with meaningful work. Like the demanufacturing work inmates conduct in order to recycle and reuse discarded goods, inmates deserve this same chance to be refurbished. US prisons should work to rehabilitate rather than punish and debilitate with hazardous work, as well as provide inmates with an outlet through which to voice complaints.

References

Atkinson, R.D. 2002. *Prison Labor: It's More than Breaking Rocks*. Policy Report, Progressive Policy Institute. Available at http://www.ppionline.org/documents/prison_labor_502.pdf [accessed 8 November 2008].

Bair, A.P. 2007. *Prison Labor in the United States: An Economic Analysis*. New York: Routledge.

Bonds, A. 2006. Punishment and profit? The politics of prisons, poverty, and neoliberal restructuring in the rural American Northwest. *Antipode* 38(1): 174-177.

CNN. Jan. 2003. Protest slams Dell's use of prison labor. [Online]. Available at http://www.cnn.com/2003/TECH/biztech/01/10/dell.protest.ap/index.html [accessed 11 November 2010].

Davis, S. and T. Smith. 2003. *Corporate Strategies for Electronics Recycling: A Tale of Two Systems*. San Jose: Publication of the Silicon Valley Toxics Coalition, Computer TakeBack Campaign.

Dayaneni, G. and A. Shuman. 2007. Toxic sentence: Captive labor and electronic waste. [Online]. Available at http://urbanhabitat.org/node/857 [accessed 11 November 2009].

Dikeç, M. 2005. Space, politics, and the political. *Environment and Planning D: Society and Space* 23(2): 171-188.

Federal Prison Industries, Inc. Annual Report 2008. 2009. Rep. Petersburg: Federal Correctional Institution.

Fouche, G. Oct. 19, 2009. *Where convicts lead the good life*. [Online]. Available at http://www.globalpost.com/dispatch/europe/091017/norway-open-prison [accessed 7 March 2012].

Gold, J.R. and G. Revill (eds) 2000. *Landscapes of Defence*. New York: Prentice Hall.

Grossman, E. 2006. *High Tech Trash: Digital Devices, Hidden Toxics, and Human Health*. Washington: Island/Shearwater Books.

Grossman, E. 2005. Toxic Recycling. *The Nation* [Online 2 November] Available at http://www.thenation.com/doc/20051121/grossman [accessed 5 January 2010].

Hallett, M.A. 2006. *Private prisons in America: a critical race perspective*. Urbana and Chicago: University of Illinois.

Jackson, A.S., A. Shuman and G. Dayaneni. 2006. *Toxic Sweatshops: How UNICOR Prison Recycling Harms Workers, Communities, the Environment, and the Recycling Industry*. Rep. San Jose, CA: Center for Environmental Health, Prison Activist Resource Center, Silicon Valley Toxics Coalition, Computer TakeBack Campaign.

Lapido, D. 2001. The rise of America's prison-industrial complex. *New Left Review* 7: 109-23.

Lepawsky, J. and C. McNabb. 2009. Mapping international flows of electronic waste. *The Canadian Geographer* 54(2): 177-195.

Luard, T. 2005. China's 'reforming' work programme. *BBC* [Online 11 May] Available at http://news.bbc.co.uk/2/hi/asia-pacific/4515197.stm [accessed 28 March 2012].

Parenti, C. 1999. *Lockdown America: Police and Prisons in the Age of Crisis*. London: Verso.

Prison Industry Enhancement Certification Program (PIECP): Program brief. 2004. *Bureau of Justice Statistic Assistance*. [Online]. Available at http://www.ojp.usdoj.gov/bja/grant/piecp.html [accessed 14 December 2008].

Reid, R.L. 2004. E-recycling behind bars. *Scrap*. [Online August/September] Available at http://www.scrap.org/ArticlesArchive/2004/Sep-Oct/ERecycling Prison.htm [accessed 10 December 2009].

Rice, S. 2006. The economic motivations of prison privatization: assessing convict labor's influence on the drive to privatize prisons. PhD diss., Rutgers, The State University of New Jersey.

Ryan, M. and T. Ward. 1989. *Privatization and the Penal System: The American Experience and the Debate in Britain*. New York: St. Martin's Press.

Sample, B. 2009. Prisoners exposed to toxic dust at UNICOR recycling factories. *Prison Legal News* 20(1): 1-5.

Shuman, A. 2007. To get stuff and sell it for as much as we can get: Federal prison industries and electronics recycling. *Prison Legal News* 18(3): 1-10.

Skillings, J. 2002. Sending old PCs up the river. *CNET News* [Online 14 February 14] Available at http:// news.cnet.com/Sending-old-PCs-up-the-river/2100-1040_3-838011.html [accessed 12 February 2010].

Stade, K. 2005. Computer recycling exposes prison staff and inmates to toxics: Prison industry safety concerns censored from OSHA investigation. *Public Employees for Environmental Responsibility* [Online 7 March] Available at http://www.peer.org/news/news_id.php?row_ id= 490 [accessed 10 January 2010].

Stephan, J. 2008. *Census of State and Federal Correctional Facilities, 2005.* United States. U.S. Department of Justice. Office of Justice Programs. Bureau of Justice Statistics. Available at http://www.ojp.usdoj.gov/bjs/pub/pdf/csfcf05.pdf [accessed 8 November 2009].

Swyngedouw, E. 2006. Circulations and metabolisms: (hybrid) natures and (cyborg) cities. *Science as Culture* 15: 105-21.

UNICOR Online. Top ten reasons to buy from UNICOR. Federal prison industries. [Online] Available at http://www.unicor.gov/about/faqs/top_ten/index.cfm [accessed 14 January 2010].

UNICOR Online. Recycling Customer List. Federal Prison Industries. [Online]. Available at http://www.unicor.gov/recycling/customerlist.cfm [accessed 7 October 2009].

UNICOR Online. Recycling. Federal prison industries. [Online]. Available at http://www.unicor.gov/recycling/ [accessed 14 January 2010].

UNICOR Online. Factory contact list map. Federal prison industries. [Online]. Available at http://www.unicor.gov/recycling/mapfront.cfm?navlocation=FactoryContacts [accessed 14 January 2010].

UNICOR Online. Recycling facts. Federal prison industries. [Online]. Available at http://www.unicor.gov/recycling/facts.cfm [accessed 12 November 2009].

U.S. Chamber of Commerce. 2005. Statement on 'Federal Prison Industries Contracting', prepared for the House Committee on the Judiciary, Subcommittee on Crime, Terrorism, and Homeland Security. [Online]. Available at http://www.uschamber.com/issues/index/govt contracting/fpi.htm [accessed 15 November 2009].

Vanderpool, T. 2006a. UNICOR unrest. *Tucson Weekly* [Online 4 May] Available at http://www.tucsonweekly.com/tucson/unicor-unrest/Content?oid=1084002 [accessed 9 December 2009].

Vanderpool, T. 2006b. Safety shell game. *Tucson Weekly* [Online 27 April] Available at 2009.http://www.tucsonweekly.com/tucson/safety-shellgame/Content?oid=1083935 [accessed 5 January 2010].

Wheaton, S. 2010. Inmates in Georgia prisons use contraband phones to coordinate protest. *New York Times* [Online 12 December] Available at http://www.nytimes.com/2010/12/13/us/13prison.html [accessed 15 March 2012].

Chapter 8
Liminal TransCarceral Space: Prison Transportation for Women in the Russian Federation

Dominique Moran, Laura Piacentini and Judith Pallot

Within the new and vibrant sub-discipline of 'carceral geography' (Moran et al. 2012, Philo 2012), a rich vein of work has emerged around the nature of carceral spaces and experiences inside and outside of them, which stress the importance of considering prisoner agency in the context of carceral control. By engaging directly with the concept of panopticism to examine a prison for women in South Africa, Dirsuweit (1999) showed that rather than being rendered 'docile', prisoner resistance to omni-disciplinary control was expressed through reclamation of culturally defined prison space. In New Mexico, van Hoven and Sibley (2008: 1016) showed that the regulation of prison space and the docility of bodies is also contested, with 'spaces … produced and reproduced on a daily basis', and inmates making 'their own spaces, material and imagined' (Sibley and van Hoven 2009: 205). This chapter extends the discourse of prisoner agency by shifting the focus away from conventionally enclosed 'spaces' of incarceration, and by considering instead the experience of incarceration outside the prison, specifically the example of prison transportation; and in so doing calling into question the binary distinction between inside and outside. In so doing it resonates with the work of Gill on Electronic Monitoring (this volume) in highlighting this problematic distinction.

The chapter opens by considering Erving Goffman's (1961) theorization of the 'total institution', critiqued elsewhere in this volume, and summarizing the critical engagements with this conceptualization of the prison, in particular the work of Farrington (1992) which addresses both the practical and the ideological mismatches between Goffman's theory and really-existing prisons, and of Baer and Ravneberg (2008) who draw attention to the spatial indistinction between the 'inside' and 'outside' of prisons as commonly understood. It extends this critique by drawing on the work of Allspach (2010) to consider the ways in which the carceral regime of the prison might be considered to reach beyond its assumed spatial boundaries, and by drawing upon geographical conceptualizations of 'liminality' to engage with women prisoners' experience of penal transportation in the Russian Federation. Having first provided some context for the empirical material collected, the chapter then presents the testimony of women prisoners recalling their experience of prison transportation or *etap*, and argues that for

these women, *etap* represents a transcarceral space of liminal transformation, which operates to blur the boundary between the 'outside' and 'inside' of the 'total institution.

Critiques of the 'total institution'

In his theorization of the 'total institution', Goffman (1961) defines 'a place of residence and work where a large number of like-situated individuals, cut off from the wider society for an appreciable length of time, together lead an enclosed, formally administered round of life' (Goffman 1961: 11). Since its inception, this terminology has been applied to an extraordinarily diverse range of circumstances and contexts, such as psychiatric units (Skorpen et al. 2008), the home (Noga 1991), homes for the elderly (e.g. Mali 2008), the mass media (Altheide 1991), the military and the police (Rosenbloom 2011), and sport (Cavalier 2011). While the concept has thereby been shown to have considerable transferability and utility, its appropriateness as a means of understanding the types of institution in relation to which it was initially developed has been widely critiqued. In this chapter, the applicability of the 'total institution' thesis to the institution of the prison is called into question; in this context, critics point out the notable disjunctures between the theory and the actuality of imprisonment. Farrington (1992: 6), for example, argues in relation to the US prison system that the 'total institution' thesis is 'in fact, fairly inaccurate as a portrayal of the structure and functioning of the … correctional institution' in that the modern prison 'is not as completely or effectively "cut off from wider society" as Goffman's description might lead us to believe'. Although Farrington (1992) addressed his critique explicitly within this geographical context, many of his observations are generically applicable to prison systems more generally, and they bear some exploration here.

At the heart of Farrington's (1992) argument is an understanding of prison institutions as having a relatively stable and ongoing network of transactions, exchanges and relationships which connect and bind them to their immediate host community and to society more generally (ibid.: 7). Although at the time of writing, Farrington (1992) observed that relatively little research had explored these connections, such as the relationships between prisons and their host communities, the process of prison siting, and the relationships between criminal offenders and the society from which they have come, in the intervening years these topics have come more clearly into view in academic scholarship, and particularly in carceral geography, which has considered precisely these issues (e.g. Che 2005, Glasmeier and Farrigan 2007, Bonds 2009, Moran et al. 2012, Pallot 2007, and McWatters – this volume). Most recently, in response to what Wacquant (2011: 3) has described as 'a brutal swing from the social to the penal management of poverty' particularly in the United States, with a 'punitive revamping' of public policy tackling urban marginality through punitive containment, and establishing a 'single carceral continuum' between the ghetto and the prison (Wacquant 2000:

384), the relationship between prisoners and wider society has been a particular focus of study (e.g. Peck 2003, Peck and Theodore 2009).

Farrington (1992: 7) essentially rejects Goffman's notion of the prison as a 'total institution' in favour of a theoretical conception of 'a "not-so-total" institution, enclosed within an identifiable-yet-permeable membrane of structures, mechanisms and policies, all of which maintain, at most, a selective and imperfect degree of separation between what exists inside of and what lies beyond prison walls'. Subsequent studies support this interpretation, with, for example, Baumer et al. (2009) describing prisons becoming 'porous' through the practice of prisoner home visits, and Hartman (2000) discussing the restriction of prisoner access to the internet in the language of 'walls and firewalls'. Although Farrington (1992) identifies 'points of interpenetration' through which the prison and wider society intrude into and intersect with one another, his critique stops short of that of Baer and Ravneberg (2008) who problematize the conceptualization of a binary distinction between 'inside' and 'outside'. Instead they posit the concept of heterotopia, viewing prisons as 'heterotopic spaces outside of and different from other spaces, but still inside the general social order' (Baer and Ravneberg 2008: 214), which they argue renders problematic the separation of inside from outside.

In their critique, Baer and Ravneberg (2008: 214) stress that further understandings of the nature of the prison, whether as a total institution or otherwise, should go beyond the perspective of the external observer, to incorporate their experience by prisoners, whose perceptions of the 'inside/outside distinctions and indistinctions [may] take on different complexities and subtleties'. An example of exactly this kind of work is Allspach's study (2010), in which by exploring the experiences of women released from federal prisons in Canada, she demonstrates that liberal 'welfarist' ideals embedded in neo-liberal reforms facilitate a network of social control of these women, forming carceral spaces beyond prison walls which perpetuate and exacerbate their marginality after release from prison. These women spoke of their post-release experiences as 'socio-economic spatial re-confinements' (Allspach 2010: 720), as they negotiated halfway-houses in deprived neighbourhoods, surveilled by closed-circuit television, and were micro-managed and surveilled by parole officers and halfway-house staff; experiences reminiscent of the surveilled institution of the prison, and which both reduced their agency and independence, and restricted their social contacts.

Seeing this work through the lens of critiques of the 'total institution', it would seem that the prison wall is permeable not only in that it permits the interpenetration of both material things (people, supplies) and less tangible things (ideas, the internet, emotional attachments), but that the 'carceral' itself is not restricted to the space contained by the permeable wall of the prison; rather that it is transported outside of the prison through the continued control of released prisoners across space, to take form elsewhere, in 'transcarceral' spaces. Transcarceral spaces may also exist alongside and perhaps also in combination with, an embodied sense of the 'carceral' which is similarly mobile beyond the prison wall through the relocation of released prisoners. In this chapter, however, we want to explore

the carceral space of *etap*, prison transportation in Russia; experienced *before* imprisonment, and existing beyond the assumed 'boundaries' of the prison, yet operating to extend the carceral reach of the penal institution. We explore the conceptualization of *etap* as a liminal space of transformation, in which convicted prisoners experience a sense of transformation from free individuals to prisoners, within the particular penal context of the Russian Federation. In the following sections of the paper we outline the theorization of liminality, and the specific research context for this paper, before presenting the extracts from interviews with women who have experienced *etap*.

Theorizing 'liminal' Space

The word liminal, from the Latin *limen,* which means boundary or threshold, has been invoked in a variety of social and cultural contexts, and by geographers has been used in combination with conceptualizations of space to describe the specific spaces of betweenness, where a metaphorical crossing of some spatial and/or temporal threshold takes place. The term was introduced first to anthropology by Van Gennep (1909, translated into English in 1960), to describe the transition from adolescence to adulthood, wherein margin or liminal is a space in which social rules are suspended because the subject no longer belongs to their old world, or to their new one – instead dwelling temporarily in 'nowhere land'. Turner (1967: 81) later developed the concept of liminality in more 'complex' societies, describing it as 'necessarily ambiguous, since this condition ... elude[s] or slip[s] through the network of classifications that normally locates states and positions in cultural space. Liminal entities are neither here nor there, they are betwixt and between the positions assigned and arrayed by law, custom, convention and ceremonial.'

Geographers have invoked this notion of liminality in a variety of contexts; exploring hotels as liminal sites of transition and transgression (Pritchard and Morgan 2006), council tenants' fora as liminal spaces between lifeworld and system (Jackson 1999), the liminal act of breastfeeding demarcating specific spaces (Mahon-Daly and Andrews 2002), liminal notions of co-existence in Australia (Howitt 2001), the street as a liminal space for prostitutes in Brazil (de Meis 2002), cyberspace as a performative liminal space for new and expectant mothers (Madge and O'Connor 2005) and most recently Gaza as a liminal territory (Bhungalia 2010). These authors highlight and problematize the 'betweenness' of the liminal, and deploy the conceptualization of the liminal as 'transformative' to frame the transitive experiences of individuals engaging with or creating these spaces. Often seen as '...intangible, elusive and obscure...', liminal spaces are 'limbo-like' and 'often beyond normal social and cultural constraints' (Preston-Whyte 2004: 350 cited in Pritchard and Morgan 2006: 764). Shields (2003: 12-13) draws particular attention to the transformation of social status allowed by liminal spaces, where 'initiates' are 'betwixt and between' life stages and where liminal spaces are bound up with ideas of becoming.

For Madge and O'Connor (2005), cyberspace allows expectant mothers to 'try out' different versions of motherhood. This work draws on Van Gennep (1960) and Turner (1967) in privileging the role of liminality as a rite of passage between one world and another, and in particular Van Gennep's (1960) description of the three stages of passage; separation from a previous life, or the 'pre-liminal'; transition, the 'liminal', and reintegration in a 'new' life, the 'post-liminal'. Turner's (1967) work focussed primarily on the liminal stage, in which he described individuals entering an unstructured egalitarian world which he termed '*communitas*', where comradeship transcends rank, age, kinship and so on, and displays an intense community spirit, in which social groups form strong bonds free from any structures which would usually constrain them. In the post-liminal (Van Gennep 1960) individuals leave *communitas* and reintegrate into their 'new' life, adopting a new social status and re-entering society in accordance with this new status.

In this chapter we intend to explore the experience of prison transportation in Russia through this framework of liminality, but before doing so, we briefly discuss the context for this research.

The Russian Prison System – Context, Data and Methodology

The data presented here were gathered through fieldwork as part of a wider project[1] within penal institutions across four Russian regions, via over 200 interviews with prison personnel and incarcerated women and girls, and also outside of these institutions, through interviews with recently released women living in three different cities in European Russia. Research was carried out between 2006 and 2010, by a team of UK and Russian colleagues permitted access to women's prisons by the Russian Federal Prison Service (*Federal'naya Sluzhba Isspolneniya Nakazaniya*, or FSIN). Access to Russian prisons was exceptionally difficult to negotiate, and the research process itself brought logistical, linguistic and ethical challenges, was always subject to institutional change and politics, and was strictly controlled by FSIN, which, like any prison administration, considered practical issues of security and institutional arrangements when allowing outsiders in. Prison research is always ethically complex, and this is particularly true in Russia, given its problematic history of prisoners' human rights. In designing the qualitative research, the usual protocols about informed consent were explained to the penal authorities and confirmation that these had been followed in obtaining volunteers for questionnaire survey and conversation was sought. However, it is almost certain, as would be the case to a greater or lesser extent in any penal context, that prisoners adjudged suitable by the prison authorities for participation in the research (on the basis of their physical, psychological and emotional state, and with concern for their health and well-being, and for the security of all concerned) were offered the opportunity to volunteer to take part, especially in the prisons where

1 ESRC award RES-062-23-0026.

the visiting western research team conducted interviews. This probably delivered a partial sample of disproportionately well-adjusted, emotionally stable respondents. To mitigate the impact of this sampling on the data generated, experienced Russian ethnographers additionally located and interviewed women who had recently been released from prison. Interviewed outside of the penal system, in their own homes or in public places of their own choosing, they reflected on and spoke more freely about, their experience of incarceration, than perhaps might women interviewed whilst in prison.

Although the data presented here pertain to female prisoners in the contemporary Russian Federation, the paper does not critically analyse their gender as a determining factor in the experience of liminality, since the gendered dimensions of their imprisonment are explored more fully elsewhere (e.g. Moran et al. 2009). In order to provide context, however, some brief information about the Russian prison system and the nature of imprisonment for women is discussed below, followed by a description of the process of *etap*, or prison transportation.

Prisons develop in context, and there are striking differences in penal interventions between countries with different historical and cultural traditions (Tonry 2001); Melossi (2001: 407) notes that '(p)unishment is deeply embedded in the national/cultural specificity of the environment which produces it'. 'Prisons are not simply institutions which (cor)respond to crime; rather, they are reflective of and mediate social, political, and cultural values, both at the level of the carceral state, and the individual prison' (Moran et al. 2009: 701). In Russia, the legacy of the Stalinist Gulag and later Soviet imprisonment practices has generated a particular penal geography (Moran 2004, Pallot 2005, 2007, Pallot et al. 2010, Moran 2011, Moran et al. 2012). Although the contemporary penal system is fundamentally different from its Soviet predecessor, imprisonment rates are still high,[2] and many penal institutions are still located in geographically peripheral locations. On 1 March 2011, 814,200 people were incarcerated in the Russian Federation, of whom 66,000 or 8 per cent were women.[3] Men and women share many aspects of Russian prison life (communal dormitories rather than cellular confinement, compulsory prison labour, different levels of privileges assigned on the basis of good behaviour, and punishment and isolation cells), but women's experience in Russia differs in the assignment of the place of imprisonment, where a sentence will be served. Of Russia's 760 correctional facilities, only 46 accommodate women, and these are unevenly distributed across space, and away from the major centres of population from which most prisoners are drawn.[4]

---

2 629/100,000 compared to 756/100,000 in the US (World Prison Population List).

3 www.fsin.su.

4 The Central Federal District, with 26 per cent of the Russian population, has just one women's prison, whereas two thirds are located in the Volga and Urals Federal Districts, which together have less than half of the population.

Almost one quarter of Russia's women's prisons are concentrated in only five[5] of its 82 regions, and women are, therefore, sent further from home to serve their sentences than are men, with implications for maintaining contact with home, family and children (Pallot 2008, Moran et al. 2009, Piacentini et al. 2009), and the ability of family and friends to visit the incarcerated.

Following arrest, all suspects are held in their nearest remand prison, usually located in the nearest urban centre, where they await trial, sometimes for several months. After trial, if convicted and sentenced to a period in custody, prisoners are allocated to the penal institution in which their sentence will be served. The allocation of institution is performed centrally by FSIN, taking into account the Penal Code (where it applies), the severity of the sentence, the appropriate penal regime (for men), and the availability of spaces in nearby institutions. For women pregnant when sentenced, the availability of spaces in institutions with mother and child units is also considered. The process of transportation itself is called '*etapirovaniya*', shortened to '*etap*' in prison jargon. The process of *etap* organized centrally by FSIN in Moscow, and prison governors and staff were reluctant to discuss it 'on the record'. However, it was established that *etap* usually takes place by train, with prisoners travelling in specially converted wagons divided up into cells, sometimes attached to the back of long-distance civilian or mail trains, frequently with no washing or toilet facilities.

Etap as Transcarceral Space

In her work on women released from federal prison in Canada, Allspach (2010) describes carceral spaces forming beyond the prison walls, in which women's marginality is exacerbated and perpetuated by networks of social control, which operate through the halfway houses in which women find themselves after release. She notes that these 'transcarceral' spaces are marked by curfews which control movement, personnel who intrude into personal affairs, closed circuit television (CCTV) surveillance, locked cupboards, restricted social contacts and restricted access to medical care. Although these spaces act, in this case, to recall the circumstances of incarceration *after* release, we argue here that the notion of transcarcerality may equally well apply to spaces experienced *before* incarceration, in which rather than extending the reach of the 'carceral' post-release, prisoners are inculcated into carceral regimes *pre*-incarceration. In the example of prison transportation in Russia, we argue here that the conditions of *etap* serve to acclimatize prisoners, both through the physical conditions of transportation, and to the overt control exerted by the penal regime.

During *etap*, the conditions in which prisoners are kept serve to acclimatize them in some ways to conditions in penal institutions. In cellular confinement in

5 Perm', Mordovia, Chuvash, Sverdlovsk, and Chelyabinsk regions have 10 prisons between them.

overcrowded train wagons, which commonly have no windows to the outside, prisoners are under direct surveillance by armed guards, and for many the conditions were both shocking and frightening. Lyudmila, interviewed during imprisonment in a juvenile prison for girls recalled, 'It was awful. We sat in a cage on the train, guards walked up and down carrying machine guns and there were dogs barking at us.' Former prisoner Svetlana described the adaptations made to the train carriages in which prisoners were transported, including the bars and the locks which kept prisoners spatially contained, and the presence of convoy guards.

> [The train carriage is] the same as a normal train, well, that is, it's the same but with bars and everything, and the locomotive is different. … Of course, guards went with us … Each of us … that is, women went separately, there was a metal grille with a lock, and bunk beds like usual on a train, but with all the bars.

On occasion the convoy personnel concealed their own identities from the prisoners: 'When we arrived [at the prison complex], all the guards were wearing masks. Just black masks, and they said that that's just how it should be. I have no idea why [they were wearing masks], it's just the rule.'

The sense of powerlessness experienced in this context was exacerbated by the transgression of certain rules pertaining to prisoner transport and the segregation of certain groups of prisoners. Former prisoner Oksana:

> They brought us to the station, it was cold, winter, and we were sitting in the freezing cold for an hour and a half waiting for the train. Then the train came, the first and the second [carriages] went to the men, and then next the women. There were 4-person coupés [compartments], but we had 10 people in them, with all their stuff in the compartment. We had kids with us too, and they were going further than us. Despite the fact that they are kids, they travelled with us. They were between 14 and 18. Although they're meant to go separately they went with us. The HIV-infected should go separately, people with TB must go separately. They went together. And who can we complain to?

She also described the ways in which prisoners were treated personally by convoy guards, who demonstrated their control not only by allocating prisoners to secure carriages, but also through violence towards certain prisoners, and by restricting access to essential facilities.

> They beat up the men, and they were condescending to the women. They didn't let us go to the toilet for 12 hours …
>
> Q: what if a person became ill?

A: 'Don't die' they'd say, 'don't die'. We don't matter. We are the dregs of society

Q: was there any humanitarian contact at all?

A: No, nothing at all.

Through a combination of factors including both the spatial context, in the form of secure train carriages containing locked cells, patrol by armed guards and dogs, and the overt and punitive control exerted by convoy personnel over transported prisoners, we would argue that *etap* serves to preface, or anticipate, the experience of incarceration for Russian prisoners, extending the penal regime of the prison into these mobile 'transcarceral' spaces which precede incarceration 'proper'. For prisoners entering these spaces, *etap* is transformative in that it prepares them in some way for what is to come. In the following section, we explore this experience by conceptualizing *etap* as a 'liminal' space.

Etap as Liminal Space

The experience of *etap* is one of betweenness – although convicted, not yet incarcerated, neither free citizen nor inmate. Turner (1967) described individuals entering liminal space experiencing an 'unstructured' world where social bonds are formed and in which the rules by which life is lived are fundamentally changed. Although *etap* is heavily structured in that it is organized and managed by the prison system, interactions take place both between transported prisoners and between prisoners and the security personnel who oversee them, and, we argue here, prisoners are prepared for imprisonment through an acclimatization process facilitated by the isolation and destabilization of *etap*, by the conversations which take place between prisoners, and by the sheer duration of the experience. After exiting this liminal space, individuals reintegrate into a 'new' life (Van Gennep 1960) as prisoners, adopting a new social status and re-entering society (within the prison) in accordance with this new status.

Resonating with Gill's (2009) depiction of asylum seekers moving around the UK's carceral estate, the liminal space of *etap* is characterized by a set of factors, which serve to isolate prisoners, separating them from their pre-liminal state and preparing them for post-liminal integration. The first of these is the withholding of information. When a prisoner receives a custodial sentence, it is commonplace for them to remain unaware of where that sentence will be served until *etap* begins, or sometimes, until arrival at the institution itself. From the perspective of friends and family, prisoners effectively 'disappear' within *etap*. Prisoners' families are not routinely informed of their whereabouts until *etap* is over and they have arrived in the institution in which a sentence will be served. Since the responsibility to inform next of kin of a prisoner's whereabouts falls to the prison management

in the receiving institution, and since the *etap* journey can take several weeks or even months, prisoners are both 'somewhere', within the system of train wagons and transit points, but 'nowhere' in that they cannot be contacted, cannot receive visits, letters or phonecalls from friends and family, and their precise whereabouts are unknown other than to convoy personnel. They appear from the outside to have 'vanished' within the system. Katya, serving a sentence in Ryazan juvenile colony described this as follows: 'No one told me where I was going. My father and brother didn't even know. My uncle found out through the police. I'd been in Ryazan' [the destination institution] for a month before my family knew where I had gone.' As Katya suggests, not only are prisoners' families unaware of their whereabouts; many prisoners also undertake the journey with no idea of their final destination. Former prisoner Tatiana described this as follows:

> In the first place, they don't tell you where they are sending you … We were in the dark about it, and when they put us on the *etap*, they didn't mention it. It's not talked about. I was told that it's a secret. You get ready, and you go on *etap*, and it's all kept a secret.

The withholding of information becomes a topic of conversation between transported prisoners, who interact during *etap*, sharing scraps of information and speculating about their destinations. Svetlana, a former prisoner who as an adult woman was transported with juvenile girls, described the confusion amongst her younger travelling companions

> Q: So you were travelling but you didn't know where to?
>
> A: I didn't know
>
> Q: Even when you were on the train?
>
> A: We talked about it in Syzran [a transit point], there the girls sat together with me and they asked what's going on, where are we going? I could only say 'I don't know where we're going'.

In some cases, convoy staff consciously actively withheld information, not only keeping prisoners 'in the dark' about their situation, but sometimes making a joke of the *etap* destination, apparently trying to frighten the prisoners by invoking both the Gulag legacy of Soviet prison camps, and the isolation of the Russian Far North. 'We were herded into these rooms, and the convoy arrived to collect us. [One of them] knocked on the door "Where are you going? To Vorkuta,[6] where the South goes on holiday".'

---

6 Vorkuta is a town in the Komi Republic, in the distant Russian North, beyond the Arctic Circle, and a destination for forced labourers in the Stalinist Gulag. During the

The social bonds which form during *etap* are not necessarily lasting friendships – prisoners may be destined for different institutions, and they frequently regard one another with a certain degree of suspicion. However, they described a certain sense of camaraderie developing during *etap*, as they were crowded together in cramped conditions, sharing meagre rations. Former prisoner Oksana:

> The rations were terrible, they barely fed us. A jar of dried potatoes, and jar of dry cereal, but no one gave us any hot water. We only got hot water in the toilet after 12 hours of travelling. We hardly ate anything, apart from what we had on us, like biscuits, or sweets. That's how it was. Just plain scary.

These interactions were facilitated by the sheer duration of *etap* for many respondents. Even if the final destination is known, prisoners have no control over or knowledge of the likely duration of the journey, since the route is plotted centrally to maximize the efficiency of transportation in general rather than to minimize journey time or disorientation for individual prisoners. For this reason, during any prisoner's journey, stops could be frequently made at other cities to pick up other prisoners, or to drop them off, frequently doubling back and taking circuitous routes. Distances travelled can be vast, and journey times of weeks or even months are commonplace. For example, Irina and Yana were both convicted in the Western Siberian city of Omsk, and sentenced to prison in Ivanovo, but they spent different lengths of time on *etap*. Irina described her experience: '[I was in transit for] two weeks – well, it was less than a month. In April I started on the *etap* and I got here in May. I arrived here on the 13th of May. [There were] three stops on the way … Kirov, Sverdlovsk [Ekaterinburg], Ivanovo.' Omsk is in Western Siberia, and although Ekaterinburg and Kirov are broadly *en route* to Ivanovo, her route involved a doubling back of 650km as the crow flies. Irina's journey was lengthy, but Yana, was in transit for almost three times as long, having travelled via an entirely different route. 'I was on the *etap* for a long time. From sometime in the middle of October, [and] I arrived here late in January, the 21st of January. I went through Chelyabinsk, then Saratov, Nizhniy Novgorod, Samara, and then here.' For both Irina and Yana, several weeks or months were spent in transit, becoming acclimatized to the penal regime. Varvara, from Bryansk region but imprisoned 750km to the north east in Vologda, was transported mainly by road vehicle, spending some time at transit points en route.

> I never saw Vologda. We were transported in closed vehicles. I was lucky, because I was driven [by car] out of Bryansk. From Bryansk, we drove for three days,

---

Soviet period, civilian inhabitants travelled on state-sponsored tours to the south of Russia in the summer to experience warm weather and sunshine. The convoy guard's joked that the prisoners were going to Vorkuta, thus reversing the direction of holiday travel.

> I think. We spent a night in a *Stolypin*,[7] and it was crazy hot, up to 70 degrees. You know what it's like, nothing to drink, nowhere to wash, for us women. Then I was in Yaroslavl prison. Then I was in transit somewhere [staying in a prison temporarily, en route] for a week, then I came to Vologda on *etap*, and I was in transit for three weeks in Vologda. I was travelling for something like a month, no, a bit less. I left on the 4th and hit the colony on the 26th.

The women interviewed conveyed a sense of the wearying experience of *etap* – that once the initial shock of the conditions had worn off, the overwhelming sense of the journey was one of tedium, and of fear of what was to come on arrival at the prison. Juvenile girls commented that: 'I was nervous and I was afraid. It was my first offence. It took two weeks to get here from Nizhny Novgorod.' 'It was frightening. You just sat there and it took a long time to arrive.'

According to Van Gennep (1960), individuals emerging from transformative liminal space reintegrate into a 'new' life, adopting a new social status and re-entering society in accordance with this status. Although *etap* served to familiarize them with some aspects of prison life, their apprehension at this sense of emergence was palpable; an awareness that what awaited them at the end of journey was a new status as prisoners. Valeria conveyed this collective sense: 'We were travelling, of course, with a sense of fear of going into the unknown, to a new place.'

The testimony of these women enables us to explore the liminal status of the spaces of *etap*; neither free nor conventionally incarcerated, and existing in tension both with the 'inside' and the 'outside' of the 'prison' proper. Returning to the debate with which this chapter opened, that of the validity or usefulness of this inside/outside distinction in relation to the 'total institution' thesis, the example of *etap* offers a number of conceptual alternatives. On the one hand, the concept of the liminal as envisaged by Turner and Van Gennep suggests that the pre- and post-liminal (the outside and inside, in the experience of transported prisoners) are clearly distinct from one another, and that the 'liminal' space of *etap* sits 'between' them. Recent scholarship in carceral geography (e.g. Baer and Ravneberg 2008, Allspach 2010) contests this view, suggesting that the inside/outside boundary is blurred, at best. If we reject the neat inside/outside distinction, then we must rethink the status of liminal space, in this case the status of spaces of *etap,* in relation to the blurred boundary between the 'outside' of 'free' life and the 'inside' of incarceration. We would argue that the adaptation processes of *etap*, drawing as they do on the visual landscapes and disciplinary practices of the prison, but situated within the civilian world of train carriages and train stations, support the notion that *etap* is a liminal transformative space; but at the same time that

---

7 In their descriptions of *etap*, prisoners and former prisoners like Varvara commonly used terminology such as '*Stolypin*', '*zek*', and '*lager*', slang words commonplace in Russian prison jargon. A Stolypin was a notorious train carriage for the transportation of prisoners, introduced in the late Tsarist era, 'zek' was a Soviet term for a prisoner, with strong Gulag connotations, and 'lager' was the Gulag name for a prison camp.

those same characteristics of *etap* highlight both the porosity between *etap* and both inside and outside, and between inside and outside themselves. If inside and outside blur into one another at an indistinct boundary, then *etap* is one example of this blurred indeterminacy, if not necessarily of 'betweenness', then perhaps of transcarceral co-presence.

Conclusion

Baer and Ravneberg (2008) appealed for geographers to challenge the total institution concept, and to do so by drawing on the experiences of prisoners traversing the 'boundaries' between inside and outside. In this chapter we have used the example of prisoner transportation in the contemporary Russian Federation to suggest that the prison wall is permeable in that 'transcarceral' spaces (Allspach 2010) exist beyond its assumed boundaries. Further, we have argued that these transcarceral spaces of transportation display some of the characteristics of liminality, as theorized by Van Gennep (1960) and Turner (1967), and of liminal space as interpreted by human geographers, in that they form a highly indistinct space, into which prisoners seem to disappear after sentencing, and out of which they emerge into the prison institution. Critically, however, this indistinct liminal space overlaps with and blurs into both the outside and inside, highlighting the porosity of the assumed boundary between them.

Philo (2012: 4) has described carceral geographies as a sub-strand of 'geographical security studies', drawing attention to its consideration of 'the spaces set aside for 'securing' – detaining, locking up/away – problematic populations of one kind or another'. However, we would argue that by considering the extent to which the 'carceral' reaches beyond the borders of the conventional 'total institution', carceral geography has the potential to engage with critiques of the 'total institution' by developing a more complex and nuanced notion of the 'carceral', as mobile, embodied and transformative.

Acknowledgements

The UK Economic and Social Research Council provided funding (RES-062-23-0026) for the research leading to this chapter. A Visiting Fellowship at the Aleksanteri Institute, University of Helsinki, provided support to Dominique Moran for the writing of the chapter.

References

Allspach, A. 2010. Landscapes of (neo)liberal control: The transcarceral spaces of federally sentenced women in Canada. *Gender, Place & Culture* 17(6): 705-723.

Altheide D. L. 1991. The mass media as a total institution. *Communications* 16(1): 63-72.

Baer, L.D. and B. Ravneberg 2008. The outside and inside in Norwegian and English prisons. *Geografiska Annaler B* 90(2): 205-216.

Baumer, E.P., I. O'Donnell and N. Hughes 2009. The porous prison: A note on the rehabilitative potential of visits home. *The Prison Journal* 89: 119-126.

Bhungalia, L. 2010. A liminal territory: Gaza, executive discretion and sanctions turned humanitarian. *GeoJournal* 75: 347-357.

Bonds, A. 2009. Discipline and devolution: Constructions of poverty, race and criminality in the politics of rural prison development. *Antipode* 41(3): 416-438.

Cavalier, E.S. 2011. Men at sport: Gay men's experiences in the sport workplace. *Journal of Homosexuality* 58(5): 626-646.

Che, D. 2005. Constructing a prison in the forest: Conflicts over nature, paradise and identity. *Annals of the Association of American Geographers* 95(4): 809-831.

De Meis, C. 2002. House and street: Narratives of identity in a liminal space among prostitutes in Brazil. *Ethos* 30(1/2): 3-24.

Dirsuweit, T. 1999. Carceral spaces in South Africa: A case study of institutional power, sexuality and transgression in a women's prison. *Geoforum* 30: 71-83.

Farrington, K. 1992. The modern prison as total institution? Public perception versus objective reality. *Crime & Delinquency* 38(6): 6-26.

Gill, N. 2009. Governmental mobility: The power effects of the movement of detained asylum seekers around Britain's detention estate. *Political Geography* 28: 186-196.

Glasmeier, A.K. and T. Farrigan 2007. The economic impacts of the prison development boom on persistently poor rural places. *International Regional Science Review* 30(3): 274-299.

Goffman, E. 1961. *Asylums: Essays on the Social Situation of Mental Patients and Other Inmates.* Garden City, NY: Anchor Books.

Hartman, K.J. 2000. Prison walls and firewalls: H.B. 2376 – Arizona denies inmate access to the internet. *Arizona State Law Journal* 32: 1423-1449.

Howitt, R. 2001. Frontiers, borders, edges: Liminal challenges to the hegemony of exclusion. *Australian Geographical Studies* 39(2): 233-245.

Jackson, N. 1999. The Council Tenants' forum: A liminal public space between lifeworld and system? *Urban Studies* 36(1): 43-58.

Madge, C. and H. O'Connor 2005. Mothers in the making? Exploring liminality in cyber/space. *Transactions of the Institute of British Geographers* 30: 83-97.

Mahon-Daly, P. and G.J. Andrews 2002. Liminality and breastfeeding: Women negotiating space and two bodies. *Health & Place* 8: 61-76.

Mali, J. 2008. Comparison of the characteristics of homes for older people in Slovenia with Goffman's concept of the total institution. *European Journal of Social Work* 11(4): 431-443.

Melossi, D. 2001. The cultural embeddedness of social control: Reflections on the comparison of Italian and North American cultures concerning punishment. *Theoretical Criminology* 5(4): 403–25.

Moran, D. 2004. Exile in the Soviet forest: 'special settlers' in northern Perm' Oblast. *Journal of Historical Geography* 30(2): 395-413.

Moran, D. 2011. Between outside and inside? Prison visiting rooms as liminal carceral spaces. *GeoJournal* [Online] Available at http://www.springerlink.com/content/p805836842373543/?MUD=MP [accessed 21 May 2012].

Moran, D., J. Pallot and L. Piacentini 2009. Lipstick, lace and longing: Constructions of femininity inside a Russian prison. *Environment and Planning D: Society and Space* 27(4): 700-720.

Moran, D., L. Piacentini and J. Pallot 2012. Disciplined mobility and carceral geography: prisoner transport in Russia. *Transactions of the Institute of British Geographers* 37 (3): 446-460.

Noga, A. 1991. Battered wives: The home as a total institution. *Violence and Victims* 6(2): 137-149.

Pallot, J. 2005. Russia's penal peripheries; Space, place and penalty in Soviet and post-Soviet Russia. *Transactions of the Institute of British Geographers* 30: 98-112.

Pallot, J. 2007. 'Gde muzh, tam zhena' (where the husband is, so is the wife): space and gender in post-Soviet patterns of penality. *Environment and Planning A*. 39: 570-589.

Pallot, J. 2008. Continuities in penal Russia: space and gender in post-Soviet geography of punishment, in T. Lahusen and P.H. Soloman Jr. (eds) *What is Soviet Now? Identities, Legacies, Memories*. LIT Verlag: Berlin, 235-254.

Pallot, J., L. Piacentini and D. Moran 2010. Patriotic discourses in Russia's penal peripheries: Remembering the Mordovian Gulag. *Europe-Asia Studies* 62(1): 1-33.

Peck, J. 2003. Geography and public policy: Mapping the penal state. *Progress in Human Geography* 27: 222-32.

Peck, J. and N Theodore 2009. Carceral Chicago: Making the ex-offender employability crisis. *International Journal of Urban and Regional Research* 32(2): 251-281.

Philo, C. 2012. Security of geography/geography of security. *Transactions of the Institute of British Geographers* 37: 1-7.

Piacentini, L., J. Pallot and D. Moran 2009. Welcome to 'Malaya Rodina' (Little Homeland): Gender, control and penal order in a Russian prison. *The Journal of Socio-Legal Studies* 18(4): 523-542.

Preston-Whyte, R. 2004. The beach as a liminal space, in C. Lew, M. Hall and A. Williams (eds) *The Blackwell's Tourism Companion*. Oxford: Blackwell, 249-259.

Pritchard, A. and N. Morgan 2005. Hotel Babylon? Exploring hotels as liminal sites of transition and transgression. *Tourism Management* 27: 762-772.

Rosenbloom, T. 2011. Traffic light compliance by civilians, soldiers and military officers. *Accident Analysis and Prevention* 43(6): 2010-2014.

Shields, R. 2003. *The Virtual.* London: Routledge.

Sibley, D. and B. Van Hoven 2009. The contamination of personal space: boundary construction in a prison environment. *Area* 41(2): 198-206.

Skorpen, A., N. Anderssen, C. Oeye and A.K. Bjelland 2008. The smoking-room as psychiatric patients' sanctuary: a place for resistance. *Journal of Psychiatric and Mental Health Nursing* 15(9): 728-736.

Tonry, M. 2001. Symbol, substance and severity in western penal policies. *Punishment and Society* 3(4): 517–36.

Turner, V. 1967. *The Ritual Process: Structure and Anti-Structure.* Harmondsworth: Penguin.

Van Gennep, A. 1960. *The Rites of Passage.* Chicago: Chicago University Press.

Van Hoven, B. and D. Sibley 2008. 'Just duck': The role of vision in the production of prison space. *Environment and Planning D: Society and Space* 26: 1001-1017.

Wacquant, L. 2000. The new 'peculiar institution': On the prison as surrogate ghetto. *Theoretical Criminology* 4: 377–89.

Wacquant, L. 2011. The wedding of workfare and prisonfare revisited. *Social Justice* 38(1–2): 1-16.

PART II
Space and Agency

Chapter 9

On Carceral Space and Agency

Yvonne Jewkes

The contributions that follow are all ostensibly about carceral space and agency but in various and imaginative ways they tease out and unpack some of the fundamental dichotomies that frame studies of imprisonment and detention: 'us' vs. 'them', inside and outside, finding privacy within a surveillance culture, association or isolation, power as opposed to powerlessness, possession and dispossession, chaos and order, and so forth. This complex assemblage of binaries is, as two of the contributors suggest, subject to multifarious – and multidisciplinary – understandings and interpretations (Milhaud and Moran). Reading the chapters from the perspective of a criminologist, this diversity was underlined. My impression is that geographers look for the positives in the carceral experience a little more than criminologists do. Yes, they highlight themes of arbitrary transfers, pointless labour, solitary confinement, indefinite detention and casual brutality, but they also present – to quote another contributor – 'dynamic' [carceral] spaces replete with temporal flows and social encounters that defy absolutist conceptions of prison as monolithic capsule of space and time' (McWatters).

It is not that criminologists and penologists have not observed agency, resistance and counter-narratives in prisons or that we do not write about themes such as autonomy, privacy and power. But we tend to emphasize the 'depth', 'weight' and 'tightness' of imprisonment (Crewe 2011), its inherent 'pains' and 'deprivations' (Sykes 1958) and society's deep cultural attachment to incarceration, manifested in a virulent 'penal populism' (Pratt 2007). The reasons why criminologists emphasize the most tragic aspects of incarceration – chronic overcrowding, suicide and self-harm, mental illness, drug addiction, high recidivism rates, low levels of literacy, and the power of the media to shape the opinions of a public infatuated with imprisonment yet deeply ignorant of its effects – probably lies not only in our epistemological roots but in our uneasy alliance with government departments. While frequently relying on them for access to prisons and sometimes for funding, it nonetheless feels incumbent upon us to point out to ministers and policy-makers whenever we have the opportunity that, on nearly all indices, prison emphatically does not 'work'.

But in these chapters, carceral geographers go beyond the pains and the problems of imprisonment and bring new insights to topics as varied as human rights and disrupted lives, notions of 'home' and the 'freedoms' that can be found within confinement, finding solitude and self within disorder, poetic testimonies and mediated spaces of punishment. A common theme across the chapters is that

of 'tactics': tactics of the body, tactics which mobilize spatial politics, tactics of challenging assaults on identity and of finding solitude in overcrowded spaces. French philosopher Michel de Certeau (1984: 29) characterizes all social relations as being a conflict between the 'strategies' of the powerful - who are assumed to be cumbersome, unimaginative and bureaucratic - and the 'tactics' of the weak - who are, by contrast, 'nimble, creative and flexible'. Institutions such as prisons and detention centres clearly constitute one of the places where the powerful construct and exercise their power, but the weak create their own spaces within those places; making them temporarily their own as they occupy and move through them.

De Certeau uses the language of warfare to describe such processes, arguing that subordinates are like guerrillas, appropriating space as a means of resistance. This metaphor seems apt in relation to the contributions here. The first two chapters remind us that there have been numerous highly publicized cases of resistance within detention centres, including riots, legal battles, suicide, self-mutilation – and hunger strike, which is the form of protest described in Chapter 10. While it might be assumed that 'hungering for freedom' is an act of literal and metaphorical diminishment, author Deirdre Conlon argues that it represents not only an attempt to counter sovereign power over 'bare life', but is a powerful agentic statement as detainees take control of their bodies within uncontrollable entanglements of space and politics. Using Agamben's work on exclusion within the modern state and Foucault's conceptualization of governmentality, Conlon discusses competing views on the extent to which, unable to resist in any material sense, asylum seekers can express agency, claim rights and establish new political practices. She concludes that counter-conducts represent a kind of freedom, a *de*subjugation that restores ontological security and has transformative power.

The progressive merging of crime control and migration control is the theme underpinning Chapter 11 in which Lauren Martin discusses the tactics which mobilize spatial politics in US immigration detention centres. In a post-9/11 climate of fear, risk and security, immigration law has become a centrepiece of anti-terrorism measures; no longer an administrative mechanism for controlling borders but a means of holding indefinitely individuals deemed 'suspicious' without having to employ the constitutional requirements that ordinarily apply to preventative detention (Aas, 2007). As Martin observes, the growing criminalization of many immigrants to the US has created a 'widening dragnet' that scoops up all those disparagingly labelled 'noncitizens'. The intertwining of penal and administrative procedures in the control of migrant populations is evident throughout Chapter 11 but Martin also presents us with evidence of detainees' challenges to the architectures of isolation, control and punishment in these environments. In particular, the visitation programs (often linked to universities or religious institutions) give detainees some autonomy and have even created alliances between detainees and staff within a general climate of disempowerment, bureaucracy and top-down manifestations of power. This in turn has made a small but significant impact on detained migrants' emotional, spiritual

and bodily experiences and to their ability to nurture a sense of 'self' within spaces that are, to all intents and purposes, prisons even if they do not bear that name.

Maintaining identity and selfhood within mortifying, dehumanizing and painful carceral environments, is a burgeoning theme in prison studies. Elsewhere I have suggested that the 'self' might best be conceptualized as the emotional 'core' that we carry from context to context (Jewkes 2002). When the public work of identity management becomes too arduous it is important to have a private place where the public mask can be put aside and one can 'be oneself'. This distinction between private sense of self and public presentation of identity is usually conceptualized in terms of 'backstage' and 'frontstage' settings respectively (Goffman 1959). For example, in the hyper-masculine environment of some men's prisons, a certain degree of controlled aggression or machismo may be required on the prison wings and in association rooms but it might be difficult to keep up this façade for long periods and individuals may need to retreat 'backstage' to private spheres in order to restore their ontological security and dissipate the tensions associated with sustaining the bodily, gestural, and verbal codes demanded of them in public (Jewkes 2002).

In Chapter 12 Olivier Milhaud and Dominique Moran explore public and private spaces in prisons and the tactics required to find solitude when living in an enforced state of 'frontstage'. Their analysis is a fascinating comparison of the impacts that French cellular confinement and Russian dormitory living have on prisoners. Many mid-eighteenth century prisons in Western Europe were constructed in response to a virulent and fatal disease known as 'gaol fever' and fear of contagion underpinned every aspect of their design and location. Inmates were compartmentalized in order to contain and control the malignant fever but rising prisoner numbers and chronic overcrowding have resulted in prisoners in France being doubled up in cells originally designed for single occupancy, or being accommodated in larger cells in groups of six to eight. In Russia, by contrast, a belief in preparing inmates for life when they return to society means that communal living in dorms is the norm. Milhaud and Moran's qualitative study illustrates that close proximity to another within the confines of a small cell can have a profoundly distressing impact, but the psychological pressure of 'batch living' (to use Goffman's phrase) in prison dormitories means that the institutional requirements for conformity and compliance are juxtaposed with the need to be engaged in constant performance (they shrewdly observe that even docility is 'performed'). Their analysis recalls to mind Sartre's observation that 'Hell is other people'. Under such circumstances, the need to 'tune out' of the prison culture and restore the sense of self, however briefly, becomes paramount, and even a few moments of silence and contemplation can have a powerfully transformative effect.

The transformative capacity of appropriating carceral space is further developed in Chapter 13, in which Julie de Dardel describes the tactics required to convert disciplinary, instrumental time into something free and creative. She describes an era of unprecedented management of spaces and bodies in the

Colombian prison system but, like Conlon in Chapter 10, she critiques Agamben's 'gloomy and univocal vision' of bare life. In a theoretically sophisticated and ethnographically vivid analysis, de Dardel observes some of the '"tricks" and daily micro-transgressions' which men and women employ to resist sovereign power in carceral spaces. Particularly striking are the descriptions of the tactics used by female inmates to counter the 'assaults on womanhood' inflicted by the institution and authorities. The joy as well as the skill of the women in Jamundí, a multi-prison complex, as they appropriate space (specifically, the space between the highest floors of the men's remand towers and the female prison blocks below), converting it into a 'chatroom' in order to communicate, flirt and seduce, is beautifully captured. Using towels skilfully coiled around their bodies in order to form alphabetical figures which spell out messages to the men below, the women are able to assert their agency, femininity and sexuality. This is but one example that de Dardel offers to illustrate the tactics of the body or, more specifically, 'tactics to clothe the body' and oppose the authorities' efforts to strip prisoners of core aspects of their identity. In de Certeau's terms, it is hard to imagine feats more nimble, creative and flexible than those practiced in the 'chatrooms' of this Colombian penitentiary.

Chapter 14 challenges the totality of the total institution in a very different way. Here Mason McWatters uses poetry written by prisoners to explore the texture of everyday life in carceral spaces. He suggests that individuals who inhabit institutional places must be understood as living and moving through complex multiplicities of experience. Personal narratives, as many of the contributions to this volume testify, demonstrate the resourcefulness of human beings even when confined in the most dehumanizing circumstances. The 'variables of experience in incarceration' described in this chapter brought to my mind Victor Hassine's powerful book, *Life Without Parole: Living and Dying in Prison Today.* Sentenced for first-degree murder in 1980, shortly after graduating from New York Law School, Hassine devoted himself to improving prison conditions and promoting prison reform through his creative writing. Tragically, after serving 28 years in various Pennsylvanian prisons, he took his own life after being denied a commutation-of-sentence hearing. However, Hassine's fiction and non-fictional writings stand as the legacy of a life confined to prison but also of a life that transcended prison. I mention Hassine's book, partly because it is a fascinating first-hand account of the architecture and lived experience of imprisonment that includes several short stories, poems and extended metaphors, and partly because it admirably verifies McWatters' statement that prison is inherently heterotopic and 'not-so-total' but can also be 'in certain instances the absolute all of one's existence'. Arguably, Hassine further reminds us of something that underpins this and all the preceding chapters, which is that even acts of suicide and self-harm in prison may be conceived as strategic and knowing acts of resistance rooted in moral and political indignation rather than as passive admissions of defeat.

Finally, Chapter 15 offers a different imagining of prison space as Jennifer Turner discusses how the spectacle of punishment operates within the media.

Focusing specifically on two programmes broadcast on British television, Turner poses several interesting questions about the mediated construction of prison inmates as essentially like 'us' or, alternatively, as undeserving 'others'. She argues that prison-themed TV programmes with heroic narratives, likeable characters or an absence of the harsh realities of incarceration have an amnesiac affect and perpetuate notions of prisons being 'soft' or 'easy' to endure. But equally, most people will never see a prison at first hand (it is probably the least visible part of the criminal justice system), and therefore the mediated prison stands in for the real thing. Consequently, and as many commentators have noted, in celebrating prison violence and encouraging voyeuristic participation among the audience, mediated prison representations have echoes of the spectacle of public executions described by Foucault (1977) in the *ancien regime*, with the audience replacing the crowd at the gallows. Perhaps even more concerning are the prisons that appear in media texts set in the future because, as Nellis observes, films such as *Fortress* (1992) and *Minority Report* (2002) may be a barometer with which to gauge the direction in which punishment is going: 'The prison of the (near) future is automated, dehumanized and secret, and it is run by sadistic and corrupt wardens working for faceless global corporations' (2006: 226). Whether there will exist opportunities for resistance in this dystopian vision of 'techno-corrections' remains to be seen, but these contributions on 'carceral space and agency' suggest that the tactics of the 'weak' within institutions designed to crush humanity, strip identity and diminish autonomy will always restore to the confined those very qualities.

References

Aas, K.F. 2007. *Globalization and Crime*. London: Sage.

Crewe, B. 2011. Depth, weight, tightness: revisiting the pains of imprisonment. *Punishment & Society* 13(5): 509-529.

De Certeau, M. 1984. *The Practice of Everyday Life*. California: University of California Press.

Foucault, M. 1977. *Discipline and Punish*. London: Allen Lane.

Goffman, E. 1959. *The Presentation of Self in Everyday Life*. New York: Anchor.

Hassine, V. 2010. *Life Without Parole: Living and Dying in Prison Today*, 5th edn. New York: Oxford University Press.

Jewkes, Y. 2002. *Captive Audience: Media, Masculinity and Power in Prisons*. Cullompton: Willan.

Nellis, M. 2006. Future punishment in American science fiction films, in P. Mason (ed.) *Criminal Visions: Media Representations of Crime and Justice*. Cullompton: Willan.

Pratt, J. 2006. *Penal Populism*. London: Routledge.

Sykes, G. 1958. *The Society of Captives: A Study of a Maximum Security Prison*. New Jersey: Princeton University Press.

Chapter 10

Hungering for Freedom: Asylum Seekers' Hunger Strikes – Rethinking Resistance as Counter-Conduct

Deirdre Conlon

Figure 10.1 The Mosney Accommodation Centre, Co. Meath, Republic of Ireland
Source: Photograph by Derek Speirs (2009). Used with permission.

Introduction

In early July 2010, a group of approximately 200 asylum seekers housed at a facility called the Mosney accommodation centre (Figure 10.1) initiated a call for a hunger strike as part of a protest against being moved to accommodation centres

in other regions in Ireland (Smyth 2010).[1] With a capacity to house up to 800 individuals, Mosney, located approximately 35 kilometres north of the capital city, Dublin, in County Meath, is Ireland's largest residential facility for asylum seekers. It was converted from its original use as a holiday camp in 2001 at a time when the number of individuals filing applications for asylum in Ireland had increased significantly over previous years.[2] In spite of its size, Mosney has been described as 'a very well-run centre' (Flynn, quoted in Casey 2010) and is viewed by some as one of the better accommodation facilities for individuals who are waiting for requests for protection as asylum claimants to be heard. The announcement, in 2010, that up to 150 individuals were to be moved from the facility was presented as a cost-saving measure that was part of government spending cuts necessitated by the state's financial crisis. A majority of those being moved were to go to an accommodation centre in a much older building located in North Dublin.

For some, the proposed move would have marked the fourth or fifth time they were relocated within Ireland in as many years. It brought the prospect of yet another upheaval of indefinite duration in already disrupted lives. This practice of regularly or irregularly moving asylum seekers around the detention and accommodation system is described as churning (Gill forthcoming) and it is common in migrant detention and incarceration across a host of geographical locations (see Hiemstra, Moran, Mitchelson this volume). The proposed churning of asylum seekers at Mosney served to illustrate how asylum seekers' rights are habitually short-changed amidst policy decisions and practices intended to govern their welfare.

At the same time, several of those who were told they would be moved had, in fact, been residing at the Mosney Centre for several years. Among those slated to move was Idriz Krasniqi; originally from Kosovo and living at Mosney for five years, he noted 'I don't want to go, I have friends here. We were given no choice' (quoted in McDonald and Hurley 2010).[3] Another protestor, Bahroz Wakachi, a Kurdish-Iranian asylum seeker, who had been waiting seven years for a decision on his asylum application and had been moved five times before being sent to Mosney, noted: 'there is great solidarity among the asylum seekers at Mosney [and] we have also had a lot of support from local people' (quoted in Casey, 2010). Thus for many individuals the centre and neighbouring community were familiar surroundings. Mosney had become a home of sorts where residents had perhaps come to understand themselves as more than asylum seekers. As I will argue in

---

1 Ireland does not have dedicated detention centres for irregular migrants, if detained, irregular migrants – including asylum seekers – are held in one of nine prison facilities. In comparative terms, asylum seekers are rarely held in prisons, nonetheless, that this occurs at all highlights the overlap between spaces of incarceration and immigrant detention.

2 Asylum applications in Ireland increased from 4,626 in 1998, to 10,938 in 2000. Applications peaked at 11,364 in 2002 but have declined to below 1998 figures since 2006.

3 Asylum seekers full names are used here as they appeared, with the permission of the interviewees, in published media accounts of the proposed hunger strike.

this chapter, by threatening to hunger strike, the asylum seekers issued a critique of policies and practices at Mosney as well as for asylum seekers more generally, even though they were not subjects of the Irish state. In doing so they engaged in what Foucault (2007) described as counter-conduct, a practice that enacts a right to question how subjects are governed, and that is wholly consonant with and immanent to the liberal government of society. For many of the protestors at the Mosney accommodation centre, the prospect of being moved to another centre was a snub of their rights as people while the call for a hunger strike represented the assertion of their capacity to engage in counter-conduct as well as their right to be political actors.

Protests that involve hunger strikes – about matters such as living conditions in detention centres, prolonged periods of waiting for refugee claims to be heard, or to defer and call attention to the plight of those facing deportation – have become a disturbingly familiar and commonplace form of protest among asylum seekers and irregular migrants in detention and detention-like settings in industrialized nations around the world. They parallel hunger strikes among incarcerated inmates in prisons and detention camps where they became widely used during the twentieth century (Scanlan et al. 2008), though, as Simanowitz (2010: 325) notes, they are 'an ancient form of protest'. These protests invite questions related to the political and social significance and impact of such disquieting and seemingly desperate practices. What circumstances prevail in carceral spaces that drive individuals to engage in such actions? What rationale can account for a person wilfully refusing food and/or fluids, which are requisite for life? How exactly do we understand agency, resistance and political acts in these settings?

Questions like these related to the logic, possibilities and limits of contestations – such as hunger strikes – among marginalized groups, are certainly not new. They have been the subject of normative inquiry and empirical investigations among a wide range of scholars (for example McGregor 2011, Tazreiter 2010, Fekete 2009, Darling 2009, Isin and Rygiel 2007, Edkins and Pin-Fat 2005, Dow 2004, Wolfram Cox and Minahan 2004). This chapter contributes to this body of scholarship by offering an analysis of hunger strikes by detained asylum seekers *vis-à-vis* Foucault's writings on governmentality. The chapter outlines scholarship that draws upon and – sometimes – critiques Agamben's conception of bare life (1998) to account for hunger strikes (Edkins and Pin-Fat 2005, Owens 2009). Without eschewing these contributions I suggest that Foucault's writings on governmentality offer a useful analytic lens through which we can come to appreciate hunger strikes – not as singular acts of resistance that overcome bare life but instead – as practices that are consonant with liberal government. To develop this line of argument, I focus on recent elucidations on Foucault's writings on counter-conducts and critical attitudes (see Conlon and Gill forthcoming, Lemke 2011, Cadman 2010) in order to argue that hunger strikes enact the right to question how one is governed, which is, at once, to recognize and realize the logic of governmentality while simultaneously refusing its form or means in a particular situation. In other words, hunger strikes exemplify a process that relies

on, yet also calls into question, the rationale and practices that are in place to guide human behaviour or conduct. As a form of protest I argue that asylum seekers' hunger strikes should not be seen as displays of agency or resistance per se but instead they should be understood as a political practice. This practice, which I refer to as 'hungering for freedom', entails freedom, subjugation, critique and desubjugation, and each of these is entangled with and immanent to the process of governmentality. This perspective extends beyond situations involving detained asylum seekers, as described here, thus, the chapter briefly considers some of the broader implications of this analysis for understanding counter-conducts as forms of protest that are exemplary, fundamental and necessary formations of political practice and subjects in liberal society.

The chapter consists of six sections. The following section presents an account of recent media interest in hunger strikes among asylum seekers. This is followed by discussion of scholarly attention to hunger strikes, paying particular attention to how these actions are accounted for and how they are sometimes cast as manifestations of heroic resistance by asylum seekers. This section also highlights how much of this work draws on and seeks to extend Agamben's work in an effort to 'retake bare life' (Owens 2009). Section four presents an overview of Foucault's work on governmentality, with an emphasis on recent discussion of counter-conducts and critical attitudes. Drawing on this work, section five outlines how Foucault's ideas offer a nuanced understanding of hunger strikes that complicates and problematizes conceptions of agency and resistance as monumental acts. In lieu of resistance, I argue that hunger strikes by detained asylum seekers enact counter-conducts and critical attitudes. These practices entail freedom, subjugation and critique, which are bound up with the conduct of conduct within a governmental framework. Furthermore, I show that as counter-conducts, hunger strikes involve desubjugation, and through this process hungering for freedom becomes a political and potentially transformative practice. The concluding section asks what lessons can be learned from this analysis that might apply to individuals and society more broadly.

Media Attention to Asylum Seekers' Hunger Strikes

Protests involving hunger strikes by asylum seekers have become more and more common in recent years. A cursory glance at media attention gives us a sense that refusals to eat – whether by declaring a fast, physically rejecting presented food items or liquids, or more visibly and physically dramatic acts that also involve self-harm such as lip sewing – are increasingly prevalent among detained asylum seekers and precariously situated migrants in different locales.[4] Recent

4 It is important to note that from a medical ethics perspective a distinction is made between hunger strikes and self-harm with the former being viewed as voluntary and for a stated political purpose. On these grounds medical intervention (e.g. force feeding) is

media reports include a February 2010 account of a five-week long hunger strike by women asylum seekers at Yarl's Wood detention and removal centre in the UK (Taylor 2010). The hunger strike was initiated by 20 women to protest the duration of their detention. It quickly garnered the support of others at the centre to the point that, over the five-week protest period, upwards of 70 detained women had participated in the protest. The Yarl's Wood centre is no stranger to hunger strikes by asylum seekers. In 2007 more than 100 women were involved in a hunger strike that focussed on poor conditions at the centre and was propelled by abusive treatment by detention centre staff (Townsend 2010). And, in 2005, Yarl's Wood was one of seven detention centres in the UK where between 57 and 112 Zimbabwean asylum seekers orchestrated a month-long hunger strike in an effort to suspend forced removals to Zimbabwe, where the risk of harm to returned migrants was heightened by post-election conflicts in the country (Allison 2010, McGregor 2009).

Hunger strikes have not only become routine among asylum seekers in the UK and Ireland, other European states have also seen increases in the practice. For instance, in September 2010, 28 Kurdish individuals in Denmark whose refugee claims had been denied, initiated a three-week long hunger strike to protest their forced deportation back to Syria, where, as a minority group, Kurds are subject to repression, imprisonment and torture by Syrian authorities (Abbas 2010a, 2010b). From late 2009 to mid-2011, there have been at least 11 hunger strikes by asylum seekers at different detention facilities as well as other locations in Greece. These include a five-week hunger strike and 77-day protest by Iranian asylum seekers that began in October 2010, a 44-day hunger strike by up to 300 migrants initiated in January 2011, and another by 29 women held at a detention centre in Ferres (http://infomobile.w2eu.net). In each of these situations, several individuals stitched their lips shut as part of their protest against intolerable conditions, inaction and injustices in government responses to asylum seekers' requests for refugee status.

To the increasing cascade of hunger strikes we must also add the infamous protests at Woomera detention centre in South Australia, where, in 2002, between 60 and 100 detained asylum seekers sewed their lips and a further 200 initiated a hunger strike to protest Australia's mandatory detention laws for irregular migrants as well as the lengthy processing times for applications for protection (see Holloway 2002). This hunger strike drew widespread media as well as political attention. The asylum seekers who sewed their lips were cast as outsiders and the practice in which they engaged was deemed culturally foreign and deeply offensive to the sensitivities of Australians (Perera 2002). Ultimately, however, media interest and popular sympathy for the protestors played a role in the eventual closure of the Woomera detention camp in 2003. From this cursory overview of a handful of recent media accounts of asylum seekers' hunger strikes we can see that

---

deemed problematic. It is beyond the scope of this chapter to explore the disturbing politics at play in these efforts to decipher what constitutes allowable versus non-permissible forms of hunger strike; see Annas (2007) for more detailed discussion of this matter.

hunger strikes are a potent form of protest that 'continue to be an effective tool in the twenty-first century' (Simanowitz 2010: 330).

Scholarly Interest in Asylum Seekers' Hunger Strikes

Causes and Effects

Media attention has been matched by scholarly interest in hunger strikes. Studies are varied in focus. Though not focussed on asylum seekers specifically, Scanlan et al. (2008) offer an empirical meta-analysis of occurrences of hunger strikes in spaces of detention and incarceration between 1906 and 2004 in order to draw conclusions about when and why hunger strikes occur. They conclude that hunger strikes emerge when all other possibilities for political action are severely curtailed and because they are an effective tactic that is used by the powerless against the state. Among the reasons hunger strikes are successful as a form of protest is their capacity to garner sympathy and consequently mobilize support from individuals and communities who are outside the detention facility. As such, emotion coupled with swells of support mean that hunger strikes, while relatively small and brief, are an effective way for seemingly powerless individuals and groups to effect change. Echoing the finding that positions of powerlessness motivate hunger strikes, Athwal and Bourne (2007), among others, examine factors that drive asylum seekers' to engage in such desperate acts. In their study of asylum seeker deaths in the UK, Athwal and Bourne observe that among those who are detained, it is commonplace to experience an overwhelming sense of isolation, arbitrary policies and inadequate provision or delayed access to medical, psychological and emotional supports. These experiences serve to amplify a sense of abandonment and intensify fears about what will happen to them if they are deported, and can thus lead asylum seekers to feel that a future of being returned to their country of origin is worth less than the risks and uncertain outcomes associated with engaging in self-harm and/or going on a hunger strike (see also McGregor 2009, Webber, 1992).

Other studies attend to the dynamics of hunger strikes and how they impact detained asylum seekers on different levels. For instance, McGregor (2009, 2011) presents a detailed report based on interviews with Zimbabwean asylum seekers in the UK, several of whom participated in the 2005 hunger strike that took place in seven different detention centres. She finds that the decision to be part of the hunger strike was empowering for some, as expressed in one interviewee's reflection 'If you really have that fear of persecution, you can say I'd rather have a dignified death here – I'll die in hospital, under my control, having chosen that – not in the way you'll be dying in Zimbabwe' (McGregor 2009: 24). Yet, for others, 'the strike was also a distressing and difficult period, and a time of a more complex politics' (ibid.). McGregor details anxieties as asylum seekers' witnessed hunger strikers becoming progressively weaker over time, feelings of being pressured to join the

protest, tensions and fears about being penalized for joining the protest and, in some cases, a lack of trust and support among protestors. Whereas the political impact of this hunger strike was successful in that it resulted in the temporary suspension of deportations to Zimbabwe, McGregor's study makes clear that the internal dynamics of these practices are fraught with insecurity and the outcomes for individuals and groups are far from certain. Thus media representations as well as political effects sometimes belie the less than unified character of these protests.

Conceptualizing Hunger Strikes as Resistance – With and Against Agamben

Moving beyond individual experiences, there is a range of scholarship that focuses on prevailing socio-political conditions – both within detention centres and in society more broadly – in an effort to account for the 'enduring power' of hunger strikes. Among these studies, several scholars draw on Agamben's conceptions of sovereign power, bare life and the state of exception (Agamben 1998, 2005) as the '*nomos* of the political space in which we are living' (Agamben 1997: 106, emphasis in the original). A detailed exegesis of Agamben's ideas is beyond the scope of this chapter. It will suffice to note that two categories – *bios* and *zoē* – are foundational to the conception and functioning of the modern state. The former refers to the capacity to engage in political life and the latter to 'humans as animals without political freedom' (Owens 2009: 572). *Zoē* is incorporated as a permanent exclusion in the modern state, that is to say, as lives or beings that, while present, are always prohibited from participating in ways that would normally be permissible for political subjects. According to Agamben, modern society is a state of exception, where being subjected to permanent exclusions and the suspension of laws and rights has become the rule. Because everyone is implicated or potentially affected by this circumstance, there is a question as to whether challenges to this state of affairs are at all possible.

Within this space, the detained asylum seeker – reduced to zoē – becomes the epitome of bare life, a perpetually absent included outsider, someone who is subjected to the laws and power of the state yet, as a subject, she or he does not have rights or recourse to the state's laws. Given this circumstance and Agamben's argument that resistance can amount to naught within this realm, hunger strikes by asylum seekers would seem to be a wholly futile exercise. Not only is this because asylum seekers are understood not to have any rights to claim in the first place, but in addition, engaging in hunger strikes merely materializes and affirms their status as outsiders. The response among politicians to the Woomera hunger strike highlights this. As Tazreiter notes 'self-harm by asylum seekers in detention centres was utilized by political leaders to reinforce the view that asylum seekers were pariahs, outside accepted values and practices' (2010: 207).

Interestingly, scholars are divided on the place of Agamben's ideas when it comes to explaining asylum seekers' hunger strikes. For some, asylum seekers' actions, particularly protests that involve lip sewing, issue a challenge to the state of exception because they embody and embrace the position of bare life (Edkins and

Pin-Fat 2005). The thinking here is that asylum seekers' actions not only expose their positions as bare life, but, in addition, they issue a refusal of the distinction between bare life and political life. In doing so, they reject the foundation upon which modern states and rights within them are based. According to Edkins and Pin-Fat, this is consistent with Agamben's perspective, where agency and contestations are possible but only if they enact a break with the existing system and materialize a situation that takes us beyond the state of exception altogether.

Other scholars propose that Agamben's ideas are useful up to a point but, in order to apprehend asylum seekers' hunger strikes, it is necessary to augment Agamben's work with insights from other theorists. For example, Owens (2009) takes issue with Agamben via a critique of Edkins and Pin-Fat's rendering of politics and resistance in asylum seekers' lip-sewing protests. Owens suggests a more nuanced approach to politics is needed in order to appreciate what is at stake in these protests, and this is possible by drawing on Hannah Arendt's views on rights and politics. In brief, Arendt rejects the notion of human rights or rights that are deemed to belong to humans in some pre-ordained fashion. Instead, rights only exist in the context of a plural community, and politics emerges through communication and exchange within this community. For Owens, the problem with Edkins and Pin-Fat's account of asylum seekers' protests is that it falls back upon a problematic notion of pre-given (human) rights. To posit that hunger strikes and lip sewing by asylum seekers are political acts that embrace bare life is to lay claim to mere physical life as a kind of human right. For Owens, what is significant and political about asylum seekers' lip sewing protests in particular is that these desperate acts call forth reaction, and, by doing so, they generate the space for dialogue and exchange that form the basis of political action. Thus, 'the politics emerging out of these actions is based on transcendence of bare life, not its celebration' (Owens 2009: 578). This perspective helps us to appreciate the significance of swells of support among the public at large for asylum seekers' hunger strikes (outlined previously, see Scanlan et al. 2008). As support develops, opportunities for exchange and communication are mobilized, and, in turn, political engagement and action are fostered.

Like Owens, Darling (2009) suggests that Agamben's insights can usefully be augmented. For Darling, Derrida's conception of hospitality offers a useful analytic lens for understanding campaigns to support failed asylum seekers. Though not focussed explicitly on hunger strikes, Darling suggests that hospitality involves an openness and willingness to question, an understanding of social dynamics as relational, and a commitment to conditional states and responses. What is interesting about this perspective is its emphasis on risk and uncertainty, which, as we will see further on, are central to understanding the character and significance of counter-conducts as a political practice.

My goal in presenting this outline of scholarship that draws on Agamben's work is not to critique it. Rather, it is to point out a distinction between these works and the perspective presented here. In this chapter, I draw on Foucault's writings – which Agamben draws on and extends (Hegarty 2010, Doyle 2009) – in order

to think through some of the ways in which individuals, such as asylum seekers, for whom political rights are severely curtailed or non-existent, claim rights and engage in political practices. The following section examines Foucault's concepts of counter-conducts and critical attitudes and offers an analysis of asylum seekers' hunger strikes from these perspectives.

Foucault, Governmentality, Counter-conducts and Critical Attitudes

In his writings Foucault traces the development and arrangement of power relations in society from control executed by a sovereign power, to disciplinary mechanisms that work to order, prohibit and occlude, and, finally, liberal government, or governmentality, which refers to more or less calculated and systematic ways of thinking and acting that propose to shape, regulate, or manage the conduct of individuals and populations (Dean 1999, see also Foucault 1991, 2007, 2008). Even though governmentality operates in ways that shape or affect individuals' actions, one of its central features is the individual's capacity to exercise freedom. In other words, individuals' actions are understood as a series of autonomous, though orchestrated, choices. Within this context, the freedom to question how one is governed is taken as part of the practice of governmentality. Foucault refers to such questions as counter-conducts or a critical attitude.[5] The objective of counter-conducts and a critical attitude is 'a different form of conduct, that is to say: wanting to be conducted differently, by other leaders [...] towards other objectives [...] and through other procedures and methods' (Foucault 2007: 194-95). Two related points are important here. First, counter-conducts and critical attitudes are 'wholly immanent and necessary to the formation of governmentality' (Cadman, 2010: 540). Thus questioning, critique, and struggles around who governs, how and why individuals or populations should be governed in particular ways, and what technologies are used in this process are embedded in and crucial to how governmentality operates. Second, because counter-conducts and critical attitudes are inherent to the process, they do not involve outright negation or rejection of governmentality; instead, they are an effort to open up – without certainty – possibilities to govern and to be governed differently.

Clearly, this perspective differs from Agamben who argues that only a radical break that 'instigates an entirely new politics' (Darling 2009: 653) would ensure that the exclusions of the state of exception are not re-instated. It also points to differences between Edkins and Pin-Fat's (2005) analysis and the one presented here. Recall their suggestion that lip sewing by asylum seekers' can be taken as a rejection of the distinction between bios and zoē and an embrace and overcoming

5 According to Foucault 'critical attitudes' emerged in secular society as a form of 'counter-conduct' or questioning about how to govern, which was present in the sixteenth century with pastoral power; as such, the two concepts are inter-related in origin and meaning. Given this, I use the terms interchangeably in this chapter.

of bare life. This rejection is viewed as an act of resistance that, in effect, establishes a new political order. In contrast, within a governmental framework, agency and resistance are problematic and counter-conducts or critical attitudes lead to a different analysis of political actions as well as the possibilities they may engender. According to Foucault, power and freedom are not understood as 'things' that are given or taken from people; instead, they are exercised through social relations and as a practice. Because of this traditional understandings of structure and agency or subjugation and resistance are moot and notions that resistance involves 'a heroic meta-subject' (Rose et al. 2006: 100) – whose actions dismantle unjust systems and replace them with alternatives – are futile. In addition, counter-conducts – in lieu of resistance – are practices that are part and parcel with governmentality; as such, they must be understood as part of a continuum rather than as singular heroic acts that bring about a radical break or a new political order.

Furthermore, counter-conducts and critical attitudes allow us to apprehend political acts and their outcomes differently. Whereas resistance is often taken as progressive or leading to some improvement in a situation, counter-conducts cannot prescribe ahead of time what the results of questioning and critique will be. As Butler notes, critique 'is a means for a future or truth that it will not know or happen to be' (2002: 215). Counter-conducts are laden with uncertainty not only for the person who engages in this practice of critique but also in the broader sense of how situations and practices of government might be affected as a result. We should thus understand counter-conducts and critical attitudes as embedded with and necessary to governmentality. They expose, problematize, and interrupt technologies of government. In this process, individuals as well as practices of government may be transformed; however, this is not a certainty and specific outcomes are not determined in advance. It is on account of their immanence in governmentality along with the fragilities and uncertainties they lay bare that counter-conducts and critical attitudes must be understood as ongoing political practices rather than as discrete or monumental acts of individual agency. In order to more fully grasp the implications of this perspective, the penultimate section considers asylum seekers' hunger strikes as instances of counter-conducts.

Asylum Seekers' Hunger Strikes as Counter-Conducts

In detention and accommodation centres, where asylum seekers' lives are clearly restricted, it may be difficult to appreciate how technologies of government, premised on individual autonomy and choice, operate and where counter-conducts might come into play. Conlon and Gill (forthcoming) demonstrate that even in the restrictive space of detention centres, asylum seekers are encouraged to become autonomous or free subjects. From the moment they come into contact with the detention estate and all the way through to the end point of being granted refugee status or the dismissal of claims and deportation, asylum seekers are guided to be autonomous, free and self-responsible individuals. Technologies of government

include discourses of choice and action that imbue the information packets that asylum seekers receive prior to being sent to residential facilities. In addition, there is the provision of educational, recreational and therapeutic classes, where asylum seekers are encouraged to become reflective entrepreneurs of the self. These freedoms, which are embedded within governmentality, also extend to opportunities to complain or critique the modes and mechanisms used to guide conduct within detention centres. These take shape as institutionalized fora for feedback, complaints and appeals as well as through unsanctioned practices such as hunger strikes. Consequently, we see that opportunities to engage in counter-conducts are part of a continuum that is entirely consonant with the freedom and autonomy that is bestowed upon individuals within a society of government.

From this we can see that hunger strikes exemplify a necessary paradox of governmentality. Asylum seekers' hunger strikes involve deploying choices and opportunities that governmentality bestows in order to critique practices of governmentality. In effect, the practice of hunger striking entails asylum seekers' subjugation as well-governed individuals, the enactment of freedoms that are extended within this realm, as well as questioning and critique, which is necessary to governmentality. But, if this is the case, a question arises as to whether or not asylum seekers' protests can be at all effective in bringing about change? Addressing this question, without resorting to romantic notions of resistance that rely on a radical break, necessitates examining how counter-conducts and critical attitudes also entail desubjugation, profound uncertainty and risk, thus why they should be understood as practices that constitute asylum seekers as ethical-political subjects.

Earlier in the chapter, I noted that in some cases asylum seekers' hunger strikes, especially those involving lip sewing, have been viewed as practices that are wholly outside the realm of cultural norms and practices (see Tazreiter 2010). On this basis, it is deemed acceptable to cast out asylum seekers as unsuitable subjects for a well-governed society. Foucault describes such situations, where individuals' actions bring the possibility of no longer being recognizable or understood, as desubjugation. And, practices that engage a critical attitude occasion the possibility of desubjugation. As Lemke observes, to critique 'means to expose one's own ontological status, it involves the danger of falling outside the established norms of recognition' (2011: 36). It is the prospect of desubjugation that throws practices and technologies of government into question, which then leads to the potential for change or transformation.

We can think about this in more concrete terms by returning to the asylum seekers' actions at the Mosney accommodation centre with which I opened this chapter. In that situation the threatened hunger strike was averted on Tuesday 6 July 2010 when the Reception and Integration Agency (RIA), which oversees residential facilities for asylum seekers in Ireland, gave an assurance that 'nobody would be forced to leave the centre' (Casey 2010) after all. Even though the asylum seekers did not embark on a hunger strike in this case, their actions enacted a form of counter-conduct that raised questions about how they were governed,

why 'like that, by these people, at this price' (Foucault 1997: 72). These questions had a number of effects. They exposed the asylum seekers as ungovernable within the current social realm. Being exposed in this way prompted questions about the effectiveness of existing technologies of government at Mosney and, perhaps, in a more general sense too. In response, the RIA conceded to the asylum seekers with a promise that nobody who did not wish to move would have to do so. Through their actions, the protestors realized a concrete outcome – nobody was moved from the centre – as well as a change to the social and political relations of government for the asylum seekers at Mosney. While their plight as asylum seekers in limbo between states remained unchanged and their lives continued to be restricted in numerous ways, they were recognized as having rights even though they were not sovereign subjects. In short, the asylum seekers' protests materialized a political space, marked by the right to question, where, through their desubjugation, a new situation became manifest and their relation as subjects as well as relations of government were also transformed.

Engaging in these practices is filled with risk: the possibility of desubjugation and, related to this, uncertainty about what the future holds. The risks and uncertainties are borne out by McGregor's research (2009, 2011), which shows that in addition to a sense of empowerment, asylum seekers' hunger strikes are filled with a sense of risk and uncertainty. In spite of these risks, asylum seekers' engage in hunger strikes in ever-increasing numbers, as this chapter outlines. This suggests that risk is an important element of these hunger strikes. Foucault notes that counter-conducts and critical attitudes involve risk because they expose situations that are unjust. This exposure makes it impossible to return to how things were before or how an individual understands and relates to her/himself. It is in this sense that counter-conducts and desubjugation are potentially transformative and in this way that they are political practices. By enacting counter-conducts asylum seekers lay claim to the right to question how they are governed despite the risks and uncertainty involved; in doing so, they become political subjects. As a result, I want to suggest that it is not lip sewing or singular acts of hunger striking that resist and defeat the extant political order. Instead, it is the perhaps more muted practice of questioning or critical attitudes, inherent within asylum seekers' hunger strikes, that effect change. In this sense counter-conducts resonate with Walters call to attend to the diverse and modest ways change occurs and for 'greater openness and sensitivity to the diverse but often relatively minor ways in which migrants are constituted, and constitute themselves […] as political subjects' (2008: 191).

Conclusion

To date there has been limited sustained attention to Foucault's writings on counter-conducts and critical attitudes, particularly in the context of detention and incarceration. This chapter contributes to such a project by arguing that asylum

seekers' hunger strikes can be understood as practices that enact hungering for freedom. That is to say, they entail counter-conducts or critical attitudes that involve autonomy, subjugation, critique and desubjugation. Each of these practices is immanent within governmentality, which hinges on the individual's capacity to exercise freedom. Through these practices, asylum seekers claim the right to question how they are governed and thus constitute themselves not merely as governed individuals but also as political subjects. This account is motivated, in part, in an attempt to consider how Foucault's writings allow development of a nuanced perspective on situations where asylum seekers are apparently reduced to subjects without political rights. In making an argument against this view, I have also been eager to dislodge renderings of hunger strikes as monumental heroic acts of resistance. Instead, consistent with a governmentality framework, I have emphasized that what is significant about hungering for freedom is the practice of questioning in and of itself. Counter-conducts, or questioning who governs, why, and how, are not discrete, singular acts of agency but instead they are contingent and continuous political practices that are embedded with the rationality and technologies of government.

In view of this, it seems fitting to ask what the implications of this understanding are, beyond the immediate context of detained asylum seekers. While Foucault would have eschewed any effort toward general, global or prescriptive statements, it is worth re-iterating that what is most significant about asylum seekers' hunger strikes are the risks and uncertainty that imbue these counter-conducts. Therefore, what can be generalised, it seems to me, is that taking a risk and asking critical questions that might alter one's relation to oneself and within society, are necessary political practices among all those who – both literally and figuratively – hunger for freedom.

References

Abbas, Z. 2010a. Failed Syrian Kurdish asylum seekers on hunger strike. *Kurd Net – Kurdish center, Kuristan news*, 16 September [Online]. Available at http://www.ekurd.net/mismas/articles/misc2010/9/kurdsworld482.html [accessed 5 March 2012].

Abbas, Z. 2010b. Syrian Kurds end hunger strike in Denmark. *Kurd Net – Kurdish center, Kuristan news*, 4 October [Online]. Available at http://www.ekurd.net/mismas/articles/misc2010/10/kurdsworld490.html [accessed 5 March 2012].

Agamben, G. 1997. The camps as the Nomos of the modern, in *Violence, Identity and Self-Determination*, edited by H. de Vries and S. Weber. Stanford: Stanford University Press, 106-118.

Agamben, G. 1998. *Homo Sacer: Sovereign Power and Bare Life*. Stanford: Stanford University Press.

Agamben, G. 2005. *State of Exception*. Translated by K. Atwell. Chicago: University of Chicago Press.

Allison, E. 2005. Detained hunger striker to continue protest. *The Guardian*, 28 June [Online]. Available at http://www.guardian.co.uk/uk/2005/jun/28/zimbabwe.politics1?INTCMP=SRCH [accessed 5 March 2012].

Annas, G. 2007. Human rights outlaws: Nuremburg, Geneva, and the global war on terror. *Boston University Law Review* 87: 427-466.

Athwal, H. and J. Bourne 2007. Driven to despair: Asylum seeker deaths in the UK. *Race and Class* 48(4): 106-114.

Butler, J. 2002. What is Critique?, in *The Political: Readings in Continental Philosophy*, edited by D. Ingram. London: Blackwell, 212-229.

Cadman, L. 2010. How (not) to be governed: Foucault, critique, and the political. *Environment and Planning D: Society and Space* 28: 539-556.

Casey, A. 2010. Mosney refugees end their hunger strike. *The Meath Chronicle*, 7 July [Online]. Available at http://www.meathchronicle.ie/news/meatheast/articles/2010/07/07/3998399-mosney-refugees-end-their--hunger-strike/ [accessed 5 March 2012].

Conlon, D. and N. Gill (forthcoming). Gagging orders: Asylum seekers and paradoxes of freedom and protest in liberal society. *Citizenship Studies*.

Darling, J. 2009. Becoming bare life: Asylum, hospitality and the politics of encampment. *Environment and Planning D: Society and Space* 27(4): 649-665.

Dean, M. 1999. *Governmentality: Power and Rule in Modern Society*. London: Sage.

Dow, M. 2004. *American Gulag: Inside U.S. Immigration Prisons*. Berkeley: University of California Press.

Doyle, S. 2009. Contesting the shape of political space: an investigation of the 'threat of Asylum' in Britain. *Refuge* 26(1): 121-134.

Edkins, J. and V. Pin-Fat 2005. Through the wire: Relations of power and relations of violence. *Millenium – Journal of International Studies* 34(1): 1-26.

Fekete, L. 2009. Europe: Crimes of solidarity. *Race and Class* 50(4): 83-97.

Foucault, M. 1991. Governmentality, in *The Foucault Effect: Studies in Governmentality*, edited by G. Burchell et al. Hemel Hempstead: Harvester Wheatsheaf, 87-104.

Foucault, L. 1997. *The Politics of Truth*. Los Angeles: Semiotext(e).

Foucault, L. 2007. *Security, Territory, Population: Lectures at the Collège de France*. Translated by G. Burchell. Edited by M. Senellart. New York: Palgrave.

Foucault, L. 2008. *The Birth of Biopolitics: Lectures at the Collège de France*. Translated by G. Burchell. Edited by M. Senellart. New York: Palgrave.

Gill, N. (forthcoming). *Peopling Immigration Control: Geographies of Governing and Activism in the British Asylum System*. Oxford: Wiley-Blackwell.

Hegarty, P. 2010. Giorgio Agamben, in *From Agamben to Zizek: Contemporary Critical Theorists*, edited by J Simons. Edinburgh: University of Edinburgh Press, 14-28.

Holloway, G. 2002. New suicide bids at Woomera. CNN World [Online]. Available at http://articles.cnn.com/2002-01-23/world/woomera.hanging_1_woomera-hunger-strikers-detainees?_s=PM:asiapcf [accessed 3 March 2012].

Isin, E. and K. Rygiel 2007. Abject spaces: Frontiers, zones, camps, in *The Logics of Biopower and the War on Terror: Living, Dying, Surviving*, edited by E. Dauphinee and C. Masters. New York: Palgrave, 181-203.

Lemke, T. 2011. Critique and experience in Foucault. *Theory, Culture, Society* 28(4): 26-48.

McDonald, D. and I. Hurley 2010. Judges to cut short holidays for asylum hearings. *Irish Independent*, 7 July [Online]. Available at http://www.independent.ie/national-news/judges-to-cut-short-holidays-for-asylum-hearings-2248436.html [accessed 5 March 2012].

McGregor, J. 2009. Narratives and legacies of detention: Zimbabwean asylum seekers' experiences in Britain. [Online: London: Migration Research Unit, Department of Geography, UCL, London Zimbabwe Association, London]. Available at www.ucl.ac.uk/news/pdf/migrationfinalreport.pdf [accessed 20 February 2012].

McGregor, J. 2011. Contestations and consequences of deportability: Hunger strikes and the political agency of non-citizens. *Citizenship Studies* 15(5): 597-611.

Owens, P. 2009. Reclaiming 'bare life'? Against Agamben on refugees. *International Relations* 23(4): 567-582.

Perera, S. 2002. What is a camp? *Borderlands e-journal* 1(1). [Online] Available at http://www.borderlandsejournal.adelaide.edu.au/vol1no1_2002/perera_camp.html [accessed 3 February 2012].

Rose, N., P. O'Malley and M. Valverde 2006. Governmentality. *Annual Review of Law and Social Science* 2: 83-104.

Scanlan, S., L. Cooper Stoll and K. Lumm 2008. Starving for change: The hunger strike and nonviolent action, 1906-2004. *Research in Social Movement, Conflicts and Change* 28: 275-323.

Simanowitz, S. 2010. The body politic: The enduring power of the hunger strike. *Contemporary Review* 292(1698): 324-331.

Smyth, J. 2010. Hundreds of asylum seekers protest over forced transfer plan from Mosney. *The Irish Times*, 6 July [Online]. Available at http://www.irishtimes.com/newspaper/ireland/2010/0706/1224274101710.html [accessed 5 March 2012].

Taylor, M. 2010. Fears for health of Yarl's Wood women in third week of hunger strike. *The Guardian*, 21 February [Online]. Available at http://www.guardian.co.uk/uk/2010/feb/21/yarls-wood-hunger-strike?INTCMP=SRCH [accessed 5 March 2012].

Tazreiter, C. 2010. Local to global activism: The movement of protect the rights of refugees and asylum seekers. *Social Movement Studies: Journal of Social, Cultural and Political Protest* 9(2): 201-214.

Townsend, M. 2010. Immigration bosses to be quizzed after asylum seekers were 'beaten' by guards. *The Guardian*, 27 February [Online]. Available at http://www.guardian.co.uk/uk/2010/feb/28/yarls-wood-assaults [accessed 3 March 2012]

Unknown. 2012. A short history of hunger strikes by refugees and migrants in Greece within 2010 and 2011 [Online]. Available at http://infomobile.w2eu.net/resistance/hunger-strikes [accessed 3 March 2012].

Walters, W. 2008. Acts of demonstration: Mapping the territory of (non-) citizenship, in *Acts of Citizenship*, edited by E. Isin and G. Nielsen. London: Zed Books, 182-206.

Webber, F. 1992. If I'm sent back, I'll die anyway. *Race and Class* 34(2): 81-85.

Wolfram Cox, J. and S. Minahan 2004. Unravelling Woomera: Lip sewing, morphology and dystopia. *Journal of Organizational Change Management* 17(3): 292-301.

Chapter 11
Getting Out and Getting In: Legal Geographies of US Immigration Detention

Lauren L. Martin

Introduction

This chapter analyses the spatial politics of US immigration detention from three angles. I first give a general overview of the US immigration enforcement system in order to situate detention in the broader regime of spatial practices used to control transboundary mobility. While immigration has long been linked to both security and crime, mandatory detention has become a key component in the process of deportation, especially under the rubric of post-9/11 counter-terrorism initiatives. Second, I outline noncitizens' pathways into and out of detention centres to show how specific forms of legal authority give US Immigration and Customs Enforcement (ICE) and Customs and Border Patrol (CBP) wide discretion to decide who to detain, where, in what conditions, and for how long. I focus on the series of discretionary decisions ICE and CBP officers make about noncitizens' individual cases, tracing detention's legal geography from the perspective of detained women and families at the T. Don Hutto Detention Center in Taylor, Texas. Third, moving from the outside in, I analyse how detention visitation programs seek to challenge detention's isolation, containment and banishment. These organizations' efforts to access detention centres – or more accurately, the detainees inside – reveal how counties, state legislators, private contractors and federal agencies claim overlapping authority to regulate detention centres. In addition, visitation programs seek to challenge detention's exclusionary function through companionship and, in doing so, focus on detained migrants' emotional, spiritual and bodily experiences. Visitation's spatial politics illuminates not only relations of power that enable detention as a spatial strategy of mobility control, but also how citizens and authorized residents confront, negotiate and contest the spatial practices of US immigration enforcement.

The chapter builds on questions raised during previous research on ICE's detention of noncitizen families (Martin 2011, 2012a, 2012b) and my participation in the formation of a visitation program at the T. Don Hutto Detention Center (Hutto hereafter) in 2010. Hutto is a 512-bed former medium security prison owned by the Corrections Corporation of America (CCA) that was retrofitted to hold families from 2006 to 2009. Half of the facility was

converted to detain adult women in 2007, and since the Obama administration's release of families in 2009, the facility has held exclusively women. Whereas my original research project explored the legal figuration of families and children as legal subjects (Martin 2011), the disciplinary functions of detention-as-deterrence policies (Martin 2012a), and family governance discourses (Martin 2012b), my experience visiting Hutto raised important questions about the spatial politics of challenging detention.[1] To further explore how visitation programs navigated detention's enclosure and isolation, I interviewed current Hutto Visitation Program participants individually and in a focus group in 2011. I knew from conversing with visitation program coordinators that other programs faced bigger institutional challenges accessing detained noncitizens. To provide a more representative picture of visitation's power geometries, I interviewed program coordinators from Sojourners' Immigration Detention Center Visitor Program (New York/New Jersey) and Detention Dialogues (San Francisco Bay Area, California).[2] Sojourners has been visiting detainees at the Elizabeth Detention Center, also owned by CCA, since 1999, while Detention Dialogues had only recently gained access to the West County Detention Center, a county-owned facility, after over a year of negotiations with ICE and county officials. I was interested not in providing sociological description of US visitation programs as a whole,[3] but rather in (1) citizens' negotiations with immigration officials and, (2) how they opened up a critical vantage point for analysing state power. While the first two sections describe noncitizens' pathways into and out of detention, the third section illuminates how immigration detention's legal geography bounds and encloses detention centres within the United States.

Enforcing Immigration: Borders, Crime and Security

Since 1889, the US federal court system has continually reaffirmed that immigration is a trade and foreign policy matter and therefore that immigration law- and policy-making belong to the legislative and the executive branches of the US federal government. Called 'the plenary doctrine', this legal manoeuvre sets immigration law apart from other constitutionally bound legal regimes,

---

1 ICE released families from Hutto in the fall of 2009, but continued to detain women there; the Berks County Family Shelter Care Center in Pennsylvania detained families until December 2011, and is currently preparing a new facility so that it can continue to hold families for ICE.

2 Ideas for and drafts of this chapter were shared with interview and focus group participants prior to publication, and I am deeply indebted to their insights.

3 Visitation Network coordinators are, however, currently in the process of collecting this kind of data, which promises to be valuable to understanding the politics of detention at different detention centres.

limiting noncitizens' constitutional protections in immigration proceedings. Until the late 1970s, however, border and immigration enforcement did not retain a high priority in public or foreign policy-making. A series of high-profile refugee arrivals from Central America and the Caribbean stoked a growing public perception in some political quarters that the US' southern borders were 'out of control'. In 1986, the Immigration Reform and Control Act instituted the first criminal penalties for immigration–related crimes into immigration law, and in the decade following, immigration law became more punitive. The 1996 Illegal Immigration Reform and Immigrant Responsibility Act (IIRIRA) mandated detention for vague--and therefore broad--categories of documented and undocumented migrants. An ever-expanding list of 'aggravated felonies' triggered mandatory detention, and the new Expedited Removal program deputized then-INS (and now ICE) officers to order certain noncitizens deported without immigration court review. As Coleman (2007a) argues, IIRIRA expanded spaces of immigration policing and shrunk judicial oversight and constitutional protections for noncitizens.

Immigration officials have relied primarily on the prison system for both detention expertise and infrastructure to meet the needs of mandatory detention. Emphasizing the deportation of 'criminal aliens', IIRIRA linked immigration status to criminality, and expanded criminal penalties (i.e., jail time) for immigration-related violations, such as carrying false identification. Thus, the growth of immigration detention as an enforcement strategy followed the discursive and institutional linking of immigration and crime, and the bulk of new detention bed-space appeared in former prisons and military barracks. Over the same period, the US prison system also grew at an unprecedented rate, fuelled by zero-tolerance and 'truth-in-sentencing' laws across the country (Austin et al. 2001, Gilmore 2007). As Teresa Miller has shown, immigration and criminal justice systems borrowed from one another, rolling back court review and privileging administrative decision-making over judicial representation (Miller 2003). Expanding the discretionary authority of intermediaries like Deportation Officers and limiting judicial discretion to release noncitizens or grant residency resonates with broader shifts in governmental strategies in other countries, as Gill (2009a) as shown in the United Kingdom. In the name of 'cutting red tape', immigration decision-making in the United States has been devolved to local jurisdictions and delegated to non-state actors, such as private prison operators. So while confinement has become a preferred method for addressing a range of public policy issues in the US, including but not limited to immigration, it also reflects broader reconfigurations of state power that increasingly embrace the proliferation of closed, opaque spaces (Gregory and Pred 2006, Paglen 2010).

Against the backdrop of 'crimmigration', post-9/11 developments in domestic security governance tied immigration to national security and territorial vulnerability (Martin 2012a). For example, Customs and Border Protection (responsible for immigration apprehensions at ports of entry, territorial borders,

and increasingly abroad, see Coleman 2007b) reframed transboundary mobility as a counter-terrorism issue:

> Prior to the terrorist attacks of 9/11, the primary focus of the Border Patrol was on illegal aliens, alien smuggling, and narcotics interdictions. ... After 9/11, it was apparent that smugglers' methods, routes, and modes of transportation are potential vulnerabilities that can be exploited by terrorists and result in terrorist weapons illegally entering the United States ... [T] he potential exists for a single individual or small group to cross the border undetected with biological or chemical weapons, weapons of mass effect, or other implements of terrorism. (Emphasis added, US Customs and Border Protection (USCBP) 2004: 4)

The border-crossers and drug traffickers of 2000 became potential conduits for terrorist operations of 2004. Building on its 1994 'Prevention through Deterrence' Southwest Border Strategy, CBP asserted 'the proper mix of assets increases the 'certainty of apprehension' of those intending to illegally cross our borders', which 'has established a deterrent effect in targeted locations'. Integrating the deterrence-through-deportation strategy into its Secure Border Initiative (SBI), the newly formed Department of Homeland Security (DHS) rolled out a bundle of border surveillance technologies (aka SBInet), increased border patrols and policy changes in 2005. DHS expanded Expedited Removal from ports of entry to zones within 100 miles of the southern or coastal borders, and to all transboundary migrants caught within 14 days of arrival. Ending 'catch and release policy', DHS targeted 'Other Than Mexicans', especially Central American migrants who could not be immediately deported and for whom INS and then DRO did not have sufficient detention space. These 'mission alignment' efforts linked border security to interior immigration enforcement and prosecution efforts. In a bold overstatement later retracted by the agency, the ICE Office of Detention and Removal's 'Operation Endgame' sought to 'deport all deportable aliens' and to 'support DHS efforts to deter illegal migration' (US Immigration and Customs Enforcement (USICE) 2003). ICE's investigative sub-agencies focused on tracking and freezing trafficking assets, which included detaining and criminally charging smugglers. While these policies cannot be said to be 'new' or specifically post-9/11, linking terrorism to immigration made unprecedented resources available to immigration authorities, and this expanded federal capacity has dramatically changed the conditions of everyday life for noncitizens in the United States.

New institutional relationships between local and federal law enforcement, on the one hand, and criminal and immigration systems, on the other hand, have created a new geography of enforcement. Through the 'Secure Communities' program all county jails will be required to share fingerprint data with immigration agencies and to notify Immigration and Customs Enforcement (ICE) of violations (USICE n.d.) by 2013. State legislatures and urban

governments have taken on immigration policy-making in response to perceived shortcomings with national immigration policy (Varsanyi 2008). Furthermore, those detained at ports of entry are increasingly charged with immigration-related crimes through the Operation Streamline program (Buentello et al. 2010). Sentenced in large groups represented by a single public defendant, these migrants increasingly serve federal prison time before being transferred to immigration detention centres and deported. This widening dragnet has led to long-term separation of families, and a heightened sense of insecurity and fear in migrant communities. In response, new and expanded immigrant rights organizations have formed to advocate for legal reform, to counter popular discourses of illegality and criminality, and to challenge specific enforcement practices (see Burridge and Loyd 2007). The interiorization of immigration enforcement, linking of crime and immigration systems, and detention and incarceration infrastructure now play central roles in the complex political geography of exclusion in the United States.

These new and expanded immigration enforcement practices made unprecedented numbers of noncitizens detainable and deportable. To address the lack of sufficient detention space, DHS opened a number of new, primarily private detention centres in the southwest United States, contracted with county jails through Inter-Governmental Services Agreements (IGSA), and added beds to existing ICE facilities. As stated above, the T. Don Hutto Detention Center is owned by the Corrections Corporation of America, but ICE contracts with Williamson County, Texas, through an IGSA, and the county subcontracts to CCA. The West County Detention Center, discussed in the third section, is a county jail, and ICE rents bed space at the facility for immigration detainees from the county. Through these kinds of arrangements, ICE expanded the number of detention beds from 19,000 in 2005 to the current level of approximately 33,400. (ICE supervises an additional 19,000 noncitizens through 'alternatives to detention' programmes.) As of September 2009, ICE owned six detention centres; of the remaining 243 detention centres, 14 are privately owned, dedicated detention facilities, and others are mixed facilities holding both convicted prisoners and civil immigration detainees (USICE 2009a). As more migrants became detainable, confinement has come to assume a central position in DHS' attempts to manage 'illegal pathways' and flows. Expanding detainability and bed space required a series of legal, administrative and infrastructural changes and concretized detention's increasingly important institutional location in immigration and border enforcement. The changing geography of enforcement is linked, as I elaborate in the next section, to spatial strategies of confinement, risk calculations and disciplinary practices that render noncitizens legible as illegal subjects.

Pathways In and Out of Detention

As a result of expanded interior immigration and border enforcement, noncitizens come into contact with ICE in a variety of ways. In addition, Arizona, Georgia and other state-level legislatures have passed legislation requiring state institutions to check immigration status. Departments of transportation, social service agencies, schools, streets and even banks have become sites of surveillance, policing and potential arrest. While these policies vary from state to state and county to county, everyday bureaucratic activities, such as drivers' license renewals, require proof of authorized residency (Shahani and Greene 2009) and cause a blurring of historically separate social, state-level and foreign, federal policy-making (Coleman 2008, Varsanyi 2008). Since 2006, ICE has also performed workplace enforcement actions and late-night visits to noncitizens' homes (popularly known as 'raids'). Noncitizens may, therefore, be rendered excludable far from the territorial border and long after entering the country. By linking with local law enforcement and a range of state institutions traditionally detached from federal law enforcement, ICE has been able to increase the potential points of contact between its agents and noncitizens (Coleman 2009).

During my research on family detention, I found that ICE detained people in a number of ways. ICE apprehended some at ports of entry, particularly those who requested asylum upon arrival. For example, a Somali mother and child crossed the Mexico-US border, and a local resident advised them not to claim asylum; the mother maintained that since her family members received asylum, she should be able to acquire asylum too. The local resident called CBP, who detained the mother and child in a temporary facility, separated them for 10 days, and then transferred them together to Hutto. Another family crossed the Mexico-US border, only to have an acquaintance call CBP, because she feared harbouring them would put her at risk. Others were picked up from their homes, and still others apprehended on city streets. In a case that became an international scandal, ICE detained a Canadian-Iranian family when their flight to Canada made an emergency landing in Puerto Rico, despite the fact that they had never intended to enter the US (Democracy Now 2007). In another instance, ICE detained a boy travelling to the US alone to join his mother. When his mother arrived to pick him up from ICE, they detained her, as well, and transferred them both to Hutto. Families' trajectories trace the geography of ICE's expanded points of contact with migrants.

Border, immigration and other homeland security agency officers sometimes work together (though often they do not, see Hiemstra, this volume), and they work for different agencies, under different mandates, and with differing authority over the deportation process. Most women and families detained at Hutto were apprehended on the Mexico-US border and moved through CBP temporary holding facilities before being transferred to ICE detention centres. CBP runs a number of temporary 'staging areas' and temporary detention centres that hold noncitizens for up to 72 hours. Those caught in the interior might pass through a

similar temporary detention centre in an ICE office or be held for ICE at a local jail. Women currently detained at Hutto may be transferred to two or three temporary facilities before CBP transfers them to ICE. Conditions vary widely between detention centres, but federal guidelines do not require temporary facilities to have beds, consistent food, warm clothing, or hygiene facilities. Migrants passing through these facilities report sleeping on the floor, in very cold conditions, with little access to restrooms and sporadic meals. Because the 72-hour restrictions are extended for weekends and state holidays, detained women can be in temporary facilities for 3 to 10 days before being transferred to a permanent facility.

After coming into contact with CBP or ICE, noncitizens face a series of categorizations that impact where they will be detained, the possibility of release, and the likelihood of their deportation. Two factors have significant bearing on noncitizens' detention trajectories: *where* and *when* they come in contact with immigration or law enforcement. Expedited Removal (ER) deputizes ICE officers to issue deportation orders to certain noncitizens without those persons seeing an immigration judge, and triggers mandatory detention; ICE typically deports ER detainees in under 30 days. This timing applies to noncitizens that request asylum just as it applies to those apprehended along the south western border. As I explain further below, ICE classifies migrants detained at a Port of Entry, however, as 'arriving aliens' and those caught inside US territory as 'illegal aliens'. These categories determine whether ICE will deem them eligible for release from detention on bond or parole.[4] Noncitizens can exit Expedited Removal by requesting a Credible Fear Interview, in which an asylum officer with Citizenship and Immigration Services will determine their fear of persecution, the first step in claiming gaining a hearing before a judge.

While the time and place of migrants' contact with immigration officials determines migrants' access to the immigration and/or asylum adjudication process, where they are detained and the duration of their detention, in practice, migrants' pathways into and out of detention are not entirely predictable. Placement decisions rest with individual ICE officers, and some regional field offices choose to release families or parents of US citizen children despite DHS' 'catch and remove' policy (see Gill 2009a, Weber 2003). While ICE leadership provides guidelines exercising discretion (e.g. USICE 2010), individual officers and field office directors are more and less sympathetic to 'vulnerable populations'. Even with seemingly clear-cut mandatory detention cases, ICE officers retain much discretion to detain, release and transfer detainees. Detained women's and families' fates are, in large part, determined by the deportation officer and immigration court judge they are assigned.

---

4 The detainee makes a monetary deposit to signify their good faith effort to pursue their immigration or asylum claim. Bond and parole amounts range from zero (rare) to $10,000, and are forfeited completely if the detainee fails to complete the legal process. This financial risk is usually borne by their families or networks, which becomes more complicated for those who do not have contacts with legal status.

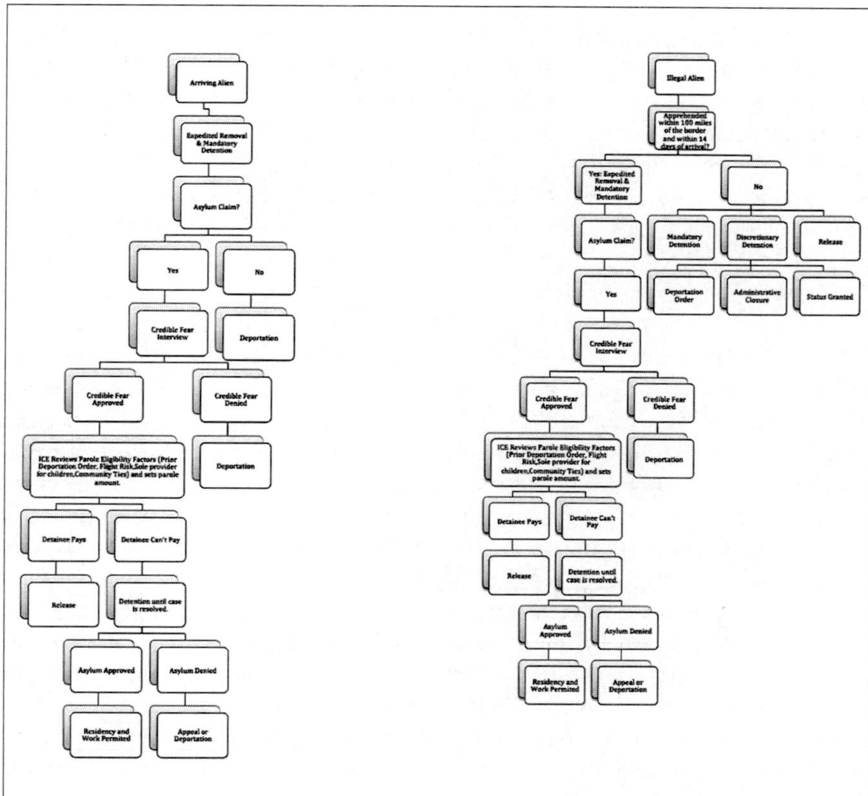

Figure 11.1 Flow chart of release, detention and deportation decisions for arriving and illegal aliens in the US immigration system

Privileging those within US territory over those outside of it (or in the process of passing into it), 'illegal aliens' can be released from detention on bond, while 'arriving aliens' can only be released on parole (see Figure 11.1). The difference between bond and parole is particularly important for asylum-seekers who claim asylum at a port of entry – that is, those who avoid undocumented entry and seek legal entrance under established humanitarian refugee law. Immigration judges determine bond eligibility and set bond amounts in noncitizens' initial hearings. ICE Deportation Officers have the discretion to set parole, however, on the basis of their evaluations of detainees' 'flight risk'. Hutto's Deportation Officers apply these evaluations in different ways, so that noncitizens' release on parole often depends on which officer they are assigned (*In re: Hutto Detention Center* Compliance Review 2 2008). Both bond and parole require substantial financial deposits, ranging from $1,500 to $10,000 per person, but getting parole requires that 'arriving aliens' find a sponsor to take responsibility for their welfare and

court appearances. For families arriving without existing family or community support, raising this kind of money can be impossible. There is then a gap between noncitizens' formal and actual opportunities for release (see Gill 2009a, 2009b and White 2002 for similar findings in the UK). The asylum process takes from six to 12 months – longer if the case is appealed – so prohibitively expensive parole amounts guarantee long detention stays, despite a Deportation Officers' evaluation of a families' low flight risk. This differentiation affords noncitizens present in US territory more procedural rights than those who have never been admitted, and this territorial hierarchy has long been considered precedent in immigration law (Neuman 2005).

Because discretionary release is not a right, but an administrative decision made by ICE or an Immigration Judge, it is not bound by the norms of due process set out in the Constitution (*Reno v. AADC* 1999). Review and relief decisions cannot be appealed to a higher or external body. In the past, federal courts adjudicated appeals, but Congress limited these appeals to increase 'administrative efficiency' and move noncitizens through the deportation process more quickly. Once detained, noncitizens are assigned an ICE Deportation Officers, and these officials decide on detainees' options for release. Federal courts have ruled, on these grounds, that constitutional due process norms do not apply to relief hearings. Because noncitizens cannot appeal discretionary relief decisions on due process grounds, relief decisions are exempt from federal or appellate court review (Neuman 2005). ICE and immigration courts retain, therefore, discretion over noncitizens' pathways through the immigration system. Though distinct, these two forms of discretionary authority resonate and work together to limit the ability of noncitizens to appeal for relief from detention and deportation.

In addition, these different legal regimes work to continually defer liability for detention conditions. In a 2007 lawsuit against the detention of families at Hutto (*In re: Hutto Detention Center* 2007), the judge found that the former medium-security prison's architecture and daily management qualified as 'prison-like' conditions; holding children there with or without their parents violated children's rights under a previous legal settlement (*Flores v. Meese* 1997). While DHS and ICE officials were legally liable for Hutto's conditions, any changes to the facility were performed by the Corrections Corporation of America (CCA), who owned and operated Hutto under contract with ICE and the local county. ICE was responsible for Hutto's conditions, but did not manage them; the county was responsible for fulfilling their contract to provide detention services to ICE, but did not operate the detention centre, and CCA was accountable to the conditions of its contract with the county, but not technically accountable to detained children. As I describe in more detail below, these deferrals of legal responsibility impact outsiders' access to detention centres, as counties, ICE, and contractors continually defer to each other rather than commit to specific visitation policies.

A number of legal mechanisms structure noncitizens' pathways into and out of detention. The time and place of their apprehension determines ICE's categorization as either 'arriving' or 'illegal alien' which directly impacts

noncitizens' options for release from detention. This categorization is followed by evaluations of flight risk, community ties, and the availability of a sponsor, which differentiates noncitizens according to institutionalized understandings of 'good' and 'bad' immigrants. ICE officers retain, however, discretion over noncitizens' detention, release, and transfer creating an opaque and unpredictable legal geography for noncitizens in immigration detention in the United States. One component in a larger assemblage of bordering practices, detention is a spatial practice through which transboundary migrants are contained in order to be made legible as noncitizens and either authorized or deported.

Challenging Detention

Arbitrary transfers, solitary confinement, indefinite detention stays, and detention's remoteness have elicited critical responses from diverse perspectives, from civil and international human rights to anti-racist and prison abolitionist to youth justice to employer trade associations. The geographies of policing, inter-jurisdictional law enforcement collaboration, forced mobility and family separation have animated noncitizen communities and immigrant rights advocates across the United States, who have used public demonstrations, vigils, media campaigns, legislative advocacy, community organizing and civil disobedience to challenge what are widely perceived to be repressive forms of state power. Among these tactics, each of which mobilize spatial politics in their own right, I focus on detention visitation here because it places authorized residents (citizens and documented noncitizens) in a direct relationship to detained individuals in immigration proceedings, immigration and detention centre staff and detention centre spaces. While they differ in scope and origin, detention visitation programs focus on detainees' emotional, bodily and spiritual experiences of detention. These programs must also negotiate with detention centre operators and ICE to gain and maintain access to detention centres, requiring specific strategies of engagement and (self-)representation (see also Moran 2011). Below, I explore how visitation illuminates detention's spatial politics. Visitors' access requires careful mediation with detention centre operators, ICE, and sometimes local officials. As described above, immigration detention is largely subcontracted to private companies and county governments (who may themselves subcontract to private companies), and detention centre operators have discretion over visitation policy at each detention centre. While ICE provides general guidelines nationally, visitors face different policies at each detention centre, revealing much about the uneven landscape of detention centre conditions and management. Working to gain access to detention from 'the outside', the process of visitation partially makes transparent the largely opaque spatial regulation of noncitizen populations.

Detention visitation programs have multiplied from four to 16 since 2009 and while they take different forms they build on shared principles of companionship and solidarity. In 2009, David Fraccaro, a long-time volunteer with Sojourners'

visitation program in New York City, began coordinating monthly conference calls, and the Detention Watch Network (DWN) provided visitors across the US with an email list-serve for increased communication and space on DWN's website for outreach to new programs. This national visitation network grew out of an emerging desire to address immigration detention's expansion 'in our own backyards'. Organizers and advocates sought ways to reach out to and serve detained people directly, and to challenge local manifestations of national anti-immigration policies. Through the national network, visitation programs affiliate around broad principles of companionship, accompaniment, human rights and advocacy, share best practices, and troubleshoot commonly shared barriers. Each program however develops its own approach, relationship to local detention officials, and procedures to respond to the specific arrangements at their local detention facility. While visitation programs have tended to be attached to universities and religious institutions, they are distinct from recognized religious ministries who offer services and counselling within detention centres. The visitation programs discussed here have modelled themselves on Sojourner's model of friendship, accompaniment, and witnessing and clearly position themselves against proselytizing (Detention Watch Network 2010). These programs do not offer formal legal advice, though some maintain relationships *with pro bono* legal representation and the Visitation Network encourages individual programs to collect and report information on human rights infringements.

While visitation policies differ from facility to facility, visitors face certain standard practices throughout the detention system. Detention centre operators usually confine visitors to public visitation spaces, and do not allow them into detention centres' interiors. While the T. Don Hutto Detention Center allows contact visits, most detention centres utilize prison-style visitation, in which visitors and detainees speak through Plexiglas and telephones.[5] Visitors are submitted to background checks and pass through some form of inspection, such as x-ray machines, at each detention centre. Visitors cannot bring belongings in with them, except for ID and car keys. Importantly, documentation of legal residency is required for visitation. Detention centre staff can revoke visitation privileges at any time for any reason. Thus, the process of visiting a detainee requires subjecting oneself to the spatial ordering and data collection regimes of the detention centre. Occupying a 'liminal space' between inside and outside (Moran 2011), visitation is therefore a fraught enterprise, especially for those engaged in other forms of opposition to detention and deportation. The process embodies – for visitors

5 In 2007, the ACLU and UT Law School Immigration Law Clinic sued ICE for holding families in prison-like conditions at T. Don Hutto Detention Center, located in Taylor, Texas (*In re: Hutto Detention Center* 2007). Contact visitation was a condition of the lawsuit's settlement, and the facility has maintained that policy and other reforms, championing Hutto as a 'humane residential facility' (see Martin 2011, 2012b). In 2011, however, a former guard was convicted of sexually abusing female detainees, highlighting the gap between agency policy and contractor behavior (Wolfson and Cargile 2011).

and for my purposes here – the arbitrary exercise of state power, the blurring of authority between federal immigration officers and contractors, and the ways in which detention centres regulate bodies within and beyond their boundaries (see also Coutin 2010, Mountz 2010). Below I discuss these dynamics in two visitation programs.

Organizers of the campaign to close the T. Don Hutto Family Residential Facility began to explore visitation after ICE released families from Hutto in September 2009. The former medium-security prison held families with minor children from 2006-2009, but converted half of its bedspace for 'noncriminal females' in 2007, when ICE found insufficient numbers of families to fill the beds. When families were released, the facility converted to an all-female facility and maintained most of the adjustments made to soften the facility from a prison to a 'residential facility' (see Martin 2011, 2012a, 2012b). Local immigrant rights organizers were concerned, however, that Hutto's conditions might backslide to pre-lawsuit conditions. Hutto's *pro bono* attorneys reported that most women were separated from their families, either in their home countries or in the US, and that this isolation caused them significant stress. Furthermore, organizers recognized that advocating around an adult population would require different tactics and a longer-term strategy, since adult detainees do not harness the same media sympathy that families had attracted.

Drawing on the Sojourner's model, Hutto volunteers adapted the orientation manual, developed relationships with local attorneys to find detainees who wanted visits, and began training volunteers with the support of the University of Texas Community Engagement Center. Contrary to organizers' fears, CCA maintained its visitation policies from family detention, allowing visitors to see one or two women for one hour per day, as long as the visitor presented CCA with the woman's ID number. This allowed HVP volunteers to visit without seeking permission or distinct authorization from CCA or ICE. Unlike other facilities, prior authorization is not required. In 2011, HVP volunteers visited 22 women over 41 visits. The program generally has a small group of regular volunteer visitors who meet with one-two women per visit, and try to see the same woman each visit until she is released or deported. HVP receives names of interested women from a local Legal Orientation Program, who carries a sign-up sheet to weekly Spanish-language legal rights sessions. The program is limited, therefore, by its reliance on Spanish speakers who attend the weekly program, a matter discussed often at the organization's monthly meetings. The HVP holds monthly 'report back' meetings so that visitors can troubleshoot any issues that arise; process the emotions drawn up by visiting, leaving, and very often hearing deeply troubling stories; and discuss how to better advocate for detained women.

As described above, detention centre contractors stipulate visitation policy. In the county jails that detain noncitizens in immigration proceedings, detainees are subject to the prison's visitation policy, as well. In most county jails, inmates must name and request the visitor, which requires that they be in touch with said visitor by phone or mail. Telephone rates are prohibitively expensive, and

detainees usually prefer to use calling cards to speak with family, rather than volunteer programs. (In many cases detainees do not have enough money for calling cards in the first place.) In contradistinction to facilities like Hutto, where visitors can request a visit without having to coordinate with the women ahead of time, county jail visitation requires phone access and money from detainees. This creates particular challenges for non-family visitors because it is difficult to establish an initial relationship with detainees sufficient to encourage them to make the request. As Detention Dialogues, a visitation program in California's San Francisco Bay area, coordinators noted, it is impossible for county jail detention visitation programs to 'fly under the radar' as at Hutto, because detainees and visitors had no way of contacting each other.

To gain access to the West County Detention Center, Detention Dialogues (DD) coordinators Christina Fialho and Christina Mansfield spent over a year getting ICE's and Contra Costa County's approval, with the support of Santa Clara University's Ignatian Center. At first, Fialho and Mansfield were unsure whether ICE was holding noncitizens in the Bay Area at all, as county jails do not often advertise their presence. After a county official claimed that Santa Clara County had never had a relationship with ICE, Fialho and Mansfield filed a public records request, which revealed that the county had been detaining noncitizens since the 1990s, but had ended its contract with ICE in 2010. Through conversations with attorneys, Fialho and Mansfield eventually located three county jails in the region holding ICE detainees: the West County Detention Facility in Contra Costa County, Sacramento County Jail and Yuba County Jail. They approached West County staff, who referred them to ICE for permission. ICE, in turn, referred them back to the jail's administrators, a deferral process frequently encountered by visitation programs and discussed on Visitation Network conference calls. Ultimately, jail administrators stated that they did not have the budget to support a visitation program. Fialho and Mansfield convinced jail staff that they were not requesting financial resources, but instead were providing *them* no-cost services by visiting and connecting detainees to wider service provider networks. Convinced of the benefit of free social services, the jail agreed to allow the visitation program to proceed, and has allowed DD to establish a free phone extension so that detainees can request visits without a calling card. (This arrangement is based on the existing system granting detainees access to *pro bono* legal representation.)

The phone system works intermittently, however, sometimes connecting to voicemail instead of ringing DD offices, sometimes disconnecting, and sometimes connecting successfully. Because of these infrastructural difficulties, DD has performed two volunteer orientations, but has yet to organize actual visits to the facility. The vagaries of the telephone system requires continual interaction and negotiation between DD coordinators and jail staff, so that building working relationships with staff members has become critical to the larger mission of the visitation program. According to DD's coordinators, certain staff members have become invested in the phone system to the extent that they, too, are frustrated with its problems and now advocate for the program themselves. Creating new

alliances between volunteers and staff, the visitation program disrupts clean boundaries between interest groups and, Fialho and Mansfield argue, opens up possibilities for advocating for detained noncitizens.

The differences between the Hutto Visitation Program and Detention Dialogues are stark. The program's respective facilities are governed by a different complex of private and inter-governmental contractors, with significant consequences for visitation programs. HVP, for example, spent considerably less time and energy gaining access and has no consistent communication with Hutto staff, while DD has spent upwards of a year figuring out where ICE detainees are held, how to visit them, building relationships with staff, and implementing a phone system. The process of gaining access to detention centres reveals an uneven geography of overlapping county, state and federal jurisdictions, private contracting and delegated authority. These recurring, localized difficulties have led Fialho and Mansfield to focus energies on advocating for national visitation policies in order to streamline new visitors' access. By sharing best practices and comparing strategies, previously isolated visitation programs have developed a broader analysis of detention's geographies of exclusion, which in turn has catalyzed a local-national, interscalar advocacy strategy.

Conclusion

Though the US Congress has not passed new immigration legislation since 1996, expanded financial resources and policy changes have transformed the institutional and material landscapes of immigration enforcement for noncitizens. Policing immigration status in more places, CBP and ICE draw more noncitizens into custody, and Expedited Removal's expansion triggers mandatory detention for more noncitizens. Once in the detention system, noncitizens face limited opportunities for release. While the Obama administration has committed to locating new detention centres near legal resources (USICE 2009b), ICE has contracted the vast majority of its detention centres in remote areas, where legal representation and familial support is scarce. The lack of proximity to advocacy groups in more populated areas decreases the chances for sustainable visitation programs for detainees, constituting a deviously shrewd form of spatial 'distance decay' on the part of private and state institutions. Once detained, noncitizens are subject to arbitrary transfers, often to federal districts where judges are unsympathetic to asylum or relief claims (Morawetz 2004). Detainees' categorizations as 'illegal' or 'arriving aliens' differentiate access to release options further, based on the time and place of contact with border/immigration enforcement. Thus pathways into and out of detention are determined by a combination of institutional resources, agency policy-making, officers' individual decision-making, and individuals' case details. This is a complex carceral and legal geography, in which noncitizens' rights are limited and the federal government's discretion to detain is extended in and through immigration law. Noncitizens are 'illegalized' (De Genova 2002)

through this assemblage of legal mechanisms, enforcement tactics, and discursive formations, and the spatiotemporality of detention renders them excludable 'by geographical design' (Mountz 2010). These relations of power are spatialized through the disciplining, forced mobility, and confinement of certain bodies; the production of surveillance, transportation and detention infrastructure; and the territorialization of certain forms of nationhood and state power.

This chapter has traced two different legal geographies of US immigration detention: noncitizens' pathways into and out of detention, and authorized residents' access to detainees and detention centres. Mediating the relationship between detainees and the wider population of authorized residents, these release and visitation policies reveal the high degree of discretion Department of Homeland Security agencies hold over the detention process. Decisions about individual migrants' cases are ostensibly guided by immigration legislation's admission and exclusion rules, yet ICE's wide discretion to decide who to detain, where, and when allows them to delegate detention functions to county jail and private corrections staff who would otherwise have no authority to enforce immigration law. The process of exclusion, or the inscribing of national boundaries onto mobile noncitizen bodies, unfolds through an assemblage of legal categorizations, discretionary release decisions, visitation policies and multi-jurisdictional governance relationships.

While detention disciplines noncitizens through surveillance and containment, it also works on noncitizens' relationships and communities, on visitors, on the potentially detainable, and on the deported (Coleman 2008, Coutin 2010, Hiemstra 2010 and this volume, Mountz 2010, Zilberg 2011). Detention's disciplinarity is not therefore confined to the detention centre itself, but acts upon and through wider networks of intimacy, support and forced mobility (Conlon 2010, Martin 2012a). Detention is not only 'a doing', as Gregory (2006) argues, but a spatial practice through which federal agencies mark both the difference between authorized and unauthorized bodies and the state's power to maintain that boundary. The regulation of detainees' and visitors' movement into and out of detention spatializes this bordering process, stretching the size, shape and scope of the border itself and localizing it far from the territorial limits of the nation-state.

Acknowledgments

Oliver Belcher, Christina Fialho, Elaine Cohen provided essential commentary during the writing process, and I am indebted to Christina Mansfield, David Fraccaro, Geoff Valdes, Rocío Villalobos, Bob Libal and members of the Hutto Visitation Project for their enthusiasm and ideas. The Mobilities, Borders and Identity Research Group in the Department of Geography at the University of Oulu, Finland generously supported research and writing this chapter. Any errors are mine alone.

Appendix A: Acronyms

CBP	Customs and Border Patrol
CCA	Corrections Corporation of America
CFI	Credible Fear Interview
DD	Detention Dialogues
DHS	Department of Homeland Security
DRO	Detention and Removals Office
ER	Expedited Removal Program
ICE	Immigration and Customs Enforcement
IIRIRA	Illegal Immigration Reform and Immigrant Responsibility Act
INS	Immigration and Naturalization Service
SBI	Secure Border Initiative
USCIS	US Citizenship and Immigration Service

References

Austin, J., M. Bruce, L. Carroll, P. McCall and S. Richards 2001. The use of incarceration in the United States. *Critical Criminology* 10: 17-41.

Buentello, T., S. Carswell, N. Hudson and B. Libal 2010. Operation streamline: Drowning justice and draining dollars along the Rio Grande. Grassroots Leadership. Available at http://grassrootsleadership.org/OperationStreamline/wp-content/uploads/2010/08/Operation-Streamline-Green-Paper.pdf [accessed 10 January 2012].

Burridge, A. and J. Loyd 2007. La Gran Marcha: Anti-racism and immigrant rights Southern California. *ACME: An International E-Journal for Critical Geographies* 6(1): 1-35.

Coleman, M. 2007a. A geopolitics of engagement: Neoliberalism, the war on terrorism, and the reconfiguration of US immigration enforcement. *Geopolitics* 12: 607-634.

Coleman, M. 2007b. Immigration geopolitics beyond the Mexico-US border. *Antipode* 39(1): 54-76.

Coleman, M. 2008. Between public and foreign policy: US immigration law reform and the undocumented migrant. *Urban Geography* 29(1): 4-28.

Coleman, M. 2009. What counts as the politics and practice of security and where? devolution and insecurity after 9/11. *Annals of the Association of American Geographers* 99(5): 1-10.

Conlon, D. 2010. Ties that bind: Governmentality, the state, and asylum in contemporary Ireland. *Environment and Planning D: Society and Space* 28: 95-111.

Coutin, S.B. 2010. Confined within: National territories as zones of confinement. *Political Geography* 29(4): 200-208.

De Genova, Nicholas. 2002. Migrant 'illegality' and deportability in everyday life. *Annual Review of Anthropology* 31: 419-47.

Democracy Now. 2007. 'I Want to be Free': 9-year-old Canadian citizen pleads from Texas immigration jail. [Online] February 23, 2007. Available at http://www.democracynow.org/2007/2/23/i_want_to_be_free_9 [accessed 12 January 2011].

Detention Watch Network. 2010. Visiting immigrants in U.S. detention facilities manual. Available at http://www.detentionwatchnetwork.org/sites/detentionwatchnetwork.org/files/Detention%20Visitation%20Manual.pdf [accessed 4 April 2012].

Flores v. Meese Settlement. 1997. Case No. CV 85-4544-RJK(Px). Available at http://www.aclu.org/pdfs/immigrants/flores_v_meese_agreement.pdf [accessed 17 June 2012].

Gill, N. 2009a. Presentational state power: Temporal and spatial influences over asylum sector decisionmakers. *Transactions of the Institute of British Geographers* 34: 215-233.

Gill, N. 2009b. Governmental mobility: The power effects of the movement of detained asylum seekers around Britain's detention state. *Political Geography* 28: 186-196.

Gilmore, R.W. 2007. *Golden Gulag: Prisons, Surplus, Crisis, and Opposition in Globalizing California*. Berkeley: University of California Press.

Gregory, D. 2006. The black flag: Guantánamo Bay and the space of exception. *Geografiska Annaler, Series B: Human Geography* 88 (4): 405-427.

Gregory, D. and A. Pred. 2006. *Violent Geographies: Fear, Terror and Political Violence*. London: Routledge Press.

Hiemstra, N. 2010. Immigrant 'illegality' as neoliberal governmentality in Leadville, Colorado. *Antipode* 42(1): 74-102.

In re Hutto Family Detention Center. 2007. Case No. A-07-CA-164-SS. Available at http://www.aclu.org/immigrants/detention/31504lgl20070826.html [accessed 30 August 2009].

Martin, L. 2011. The geopolitics of vulnerability: Migrant families in U.S. immigrant family detention policy. *Gender, Place, & Culture* 18(4): 477-498.

Martin, L. 2012a. Catch and return: Detention, deterrence, and discipline in US noncitizen family detention practice. *Geopolitics* 17(2): 312-334.

Martin, L. 2012b. Governing through the family: Struggles over US noncitizen family detention policy. *Environment and Planning A* 44(4): 866-888.

Miller, T. 2003. Citizenship and severity: Recent immigration reforms and the new penology. *Georgetown Immigration Law Journal* 17: 611-666.

Moran, D. 2011. Between outside and inside? Prison visiting rooms as liminal carceral spaces. *GeoJournal*. Published online December 4, 2011. http://link.springer.com/article/10.1007%2Fs10708-011-9442-6?LI=true [accessed 4/1/2013]

Morawetz, N. 2004. Detention decisions and access to Habeas Corpus for immigrants facing deportation. *Boston College Third World Law Journal* 25: 13-33.

Mountz, A. 2010. *Seeking Asylum: Human Smuggling and Bureaucracy at at the Border*. Minneapolis: University of Minnesota Press.

Neuman, G. 2005. Discretionary deportation. *Georgetown Immigration Law Journal* 20: 611-666.

Paglen, T. 2010. *Invisible: Covert Operations and Classified Landscapes.* New York: Aperture Press

Reno v. AADC 1999. 525 U.S. 471.

Shahani, A. and J. Greene 2009. *Local Democracy on ICE: Why State and Local Governments Have No Business in Federal Immigration Law Enforcement.* Brooklyn, NY: Justice Strategies.

US Customs and Border Protection. 2004. *National Border Patrol Strategy.* Department of Homeland Security: Washington, DC.

US Immigration and Customs Enforcement n.d. *Secure Communities.* [Online] Available at http://www.ice.gov/secure_communities/ [accessed 11 January 2012].

US Immigration and Customs Enforcement. 2003. *Endgame: Office of Detention and Removal Strategic Plan, 2003-2012.* US Department of Homeland Security: Washington, D.C.

US Immigration and Customs Enforcement. 2009a. *Immigration Detention Overview and Recommendations (by Dora Schriro)* Department of Homeland Security: Washington, DC.

US Immigration and Customs Enforcement. 2009b. Press Release. August 6, 2009. Washington, D.C.

US Immigration and Customs Enforcement. 2010. Memo: Civil Immigration Enforcement: Priorities for Apprehension, Detention, and Removal of Aliens. Policy No. 10072.1 On file with author.

Varsanyi, M. 2008. Immigration policing through the backdoor: City ordinances, the 'right to the City', and the exclusion of undocumented day laborers. *Urban Geography* 29(1): 29-52.

Weber, L. 2003. Down that wrong road: Discretion in decisions to detain asylum seekers arriving at UK ports. *Howard Journal of Criminal Justice* 42: 248-262.

White, A. 2002. Geographies of asylum, legal knowledge and legal practices. *Political Geography* 21(8): 1055-1073.

Wolfson, S. and E. Cargile 2011. Former guard takes plea deal for abuse: T. Don Hutto case not closed in Williamson County. KXAN.com 17 May 2011. Available at http://www.kxan.com/dpp/news/crime/Decision-day-in-T-Don-Hutto-abuse-case [accessed 11 January 2012].

Zilberg, E. 2011. *Space of Detention: The Making of a Transnational Gang Crisis between Los Angeles and San Salvador.* Durham, NC: Duke University Press.

Chapter 12
Penal Space and Privacy in French and Russian Prisons

Olivier Milhaud and Dominique Moran

Introduction

The work of Foucault (1979) has heavily influenced the work of geographers engaging with carceral space. Self-surveillance as the mechanism through which disciplinary power operates to produce 'docile' bodies has been challenged by Foucault himself through the importance of immanent resistance in any governmental framework (1991), and by all those who argue for the significance of prisoner agency; that socialization mitigates the effect of biopower, that prisoners 'perform' docility rather than interiorizing it, and who identify the mediating importance of penal space in the operation of biopower (Dirsuweit 1999, Vaz and Bruno 2003, Baer 2005, Simon 2005, Sibley and Van Hoven 2009). Based on their study of dormitory confinement in a New Mexico prison, Sibley and Van Hoven (2009) in particular call for more detailed empirical analysis of the experience of penal space by inmates, to further explore the ways in which prisoners respond to constant surveillance in the penal setting. Understandings of carceral space are key to the current development of carceral geography. However, relatively little attention has been paid to the different types of organization of space within prison walls and their effect on inmates' agency.

This chapter opens a space for discussion of the experience of penal space by investigating prisoner experiences in relation to the concept of privacy in two different penal settings – France and the Russian Federation. It first overviews the theorization of privacy within geographical scholarship, before considering the spatial organization of prisons in France and Russia. After a brief methodological discussion, empirical materials collected in the two contexts are presented and analysed. Finally it draws together some conclusions from the comparison of these two contexts.

Conceptualizing Private and Public in Penal Space

In this chapter we explore what 'privacy' might mean in the penal context. The public/private distinction is complex – it does not represent a single paired opposition, but a complex arrangement of binaries, specific to particular contexts

and subject to different interpretations and understandings. Further, although the 'public' and 'private' realms map onto and correspond to spaces in which life is lived, there is no intrinsic set of characteristics for either of these types of space. Drummond (2000: 2379) discusses the conceptualization of public and private space, summarizing the Western academic debate around these terms in which 'private space is considered to be a domestic space where social reproduction occurs more or less free from outright control by outside forces such as the state … and public space is "out there", belonging to the whole community, although regulated by prevailing social and legal norms'. This inside/outside dichotomy has been analysed and critiqued, with feminist scholars pointing out the patriarchal character of the association between masculinity and the public, and femininity and the private/domestic, in the light of the fact that the domestic can be the locus of oppression rather than freedom. With this in mind, scholarship on public spaces explores their position in mediation between public and private, masculine and feminine. For example, Jackson's (1998: 188) work on the shopping mall as a domesticated public space challenges the public/private binary and suggests that engagement with the public and the private must be 'sensitive to the socially differentiated nature of these highly contested spaces'. Mitchell (2003: 182) has strongly underlined the 'exclusionary citizenship' that is promoted in contemporary public spaces. Geographers in particular have disrupted the boundaries between public and private space, suggesting ways of destabilizing the assumed binary between them; Fenton (2005) identifies citizenship in private space; Blomley (2005) argues that public and private categorizations of space are fluid, rather than mutually exclusive and exhaustive; Allen (2006) identifies 'privatized public spaces' in Berlin, Tyndall (2010) challenges the assumed 'publicness' of shopping malls in Sydney, Australia; Kumar and Makarova (2008) argue for the 'domestication' of public space, and Staehli (2010: 72) reminds us of 'the importance of putatively private spaces of public deliberation and action'.

Within carceral space, confiscation of privacy has been described as a 'functional prerequisite' of imprisonment (Schwartz 1972: 229), and one of Sykes' 'pains of imprisonment' (1958). Foucault (1979) argued that prisoners are almost always in one another's presence, or in the sight of penal authorities, in the constantly surveilled 'panopticon'. This lack of privacy at an everyday level can include forced exposure (e.g. communal showers); forced spectatorship (exposure to others' lack of privacy); and violation of collective privacy (imposed and exposed intimacies) Schwartz (1972). Much recent discussion of privacy within criminology and prison sociology does not theorize prisoner privacy per se, rather considers it in relation to prison overcrowding, where single-cell or small unit accommodation is perceived as a means of reducing noise, constant activity and violations of personal space which can arguably lead to stress responses and to an increased likelihood of disorder and violence (Schaeffer et al. 1988, Grant and Memmott 2007, Sharkey 2010). In this chapter we argue that the nature of penal spaces has resonance beyond their connection to penal disorder; in particular we respond to Sibley and Van Hoven's (2009) commentary on the agency of prisoners

constructing material and imaginary spaces, and to their call for more detailed empirical analysis to illuminate the production of space in the carceral setting. We do this specifically by comparing the experience of penal space on the part of inmates incarcerated in two different penal settings – those of France and the Russian Federation. The following section describes, in broad-brush, the contrasts between these two penal systems in terms of their carceral estates, and the research which was carried out in each of them to generate the data presented in this chapter.

Penal Contexts and Methodologies

The penal geographies of France and of Russia differ significantly, both at the macro and the micro scale. Whereas in Russia the geography of punishment was heavily influenced by the Soviet Gulag and the location of prisons in geographically peripheral areas (see Moran et al. 2011), in France older prisons tend to be in the centre of cities, or very close to them, while new, larger prisons are often located at the urban outskirts, but still relatively close to centres of population.[1] The number of prisons in France fell sharply from the nineteenth century as the small prisons that were previously to be found in towns and small cities were closed. The penal population was therefore concentrated in large cities; half of French prisons are located in prefectures and subprefectures, i.e. the main French cities. 60 per cent of inmates are now held in urban areas with populations greater than 200,000. In other words, in contrast with Russia, most inmates in France are held in population centres, at the heart of transport networks. Although the oldest prisons, previously located in the city centres of prefectures and subprefectures have largely been closed, they have been replaced by larger prisons in the outskirts of these main cities. For the French penitentiary administration, it is important for prisons to have easy access to highways, to hospitals and to courts, and to be close to the centres of population (Milhaud 2009).

At the scale of the individual institution, the spaces of confinement also differ significantly. The architecture of French prisons promotes the separation of inmates through cellular confinement (Demonchy 1998, 2000, 2004), with the intention of dividing the carceral population into small groups moving through specific spaces at specific times, to prevent any gathering or collective organization of a large number of inmates, and echoing the concern for penal disorder noted by criminologists. Within such a spatial and temporal organization, it is very

1 Many prisons were founded during the French Revolution. They were established close to courthouses, in the centres of the main French cities. The proximity of the courthouse and prison highlighted the effectiveness of the juridical power of the State. Furthermore, in the prevailing ideology, the power of the State had to be visible throughout French territory and in each urban centre of each administrative unit (*départements*). Today, the need to be close to the courts, is outweighed in terms of prison location by the requirement for a large and relatively cheap parcel of land (Milhaud, 2009).

difficult for inmates to freely associate, especially since many inmates in *maisons d'arrêt*[2] are locked in their cells for up to 22 hours per day. Furthermore it is very difficult for an inmate to become acquainted with any appreciable proportion of the prison population, except through hearsay and rumour. In France, within prison buildings, wings are subdivided not only into different cells but also in different zones. The recommendations from the Ministry of Justice are clear: 'It is indispensable that any *maison d'arrêt* has wings as impenetrable as possible, enabling an internal division of the penal population, according to juridical criteria (first-time offenders/recidivists, criminal offence/minor offence) and personality criteria (sex, age, mental and physical health, drug addiction, dangerous behaviour)' (French Ministry of Justice 1990: 12). Hence, from an architectural point of view, French prison space is full of spatial discontinuities, walls, fences, airlocks, which act as a safety device. As one prison architect puts it, 'the barriers work as a firebreak. We compartmentalize everything. It is like containing fire'. As Salle (2011) explains, the fear of contagion is a key aspect of prison architecture in France: fear of physical contagion (disease), fear of moral contagion (vice), and fear of political contagion (revolt). Cellular accommodation has been seen as the best way to prevent these contagions and to overcome these fears.

By contrast, in the Russian system, although cellular accommodation does exist, it has been restricted to the strictest regime prisons for men, to remand prisons, and to the disciplinary sections of other types of penitentiary. Instead of cells, prisoners are commonly accommodated together in communal detachment blocks; essentially large rooms partially subdivided into two or more parts, and accommodating between 60 and 150 individuals. The accommodation of prisoners in communal or shared detachment blocks reflects the history of penal architecture; as Oleinik (2003: 49) notes, imprisonment in cells has never played a major role in Russia. As King (1994) has described, the Soviet system of corrective labour which underpinned this longstanding preference for communal confinement derives from Lenin's insistence that deprivation of liberty should not involve isolation from society. Discussing post-Soviet imprisonment, Oleinik (2003) highlights imprisonment in groups as a critical factor in the social organization of Russian prison life. Sykes has argued 'the prison is a society within a society' (1958: xii), and Oleinik draws attention to the parallels between imprisonment in

2 In the French penitentiary system, 101 of the 189 prisons are *maisons d'arrêt*, i.e. prisons operating both as remand prisons and as prisons for people serving short sentences. People who are held in *maisons d'arrêt* usually stay in their cells during the day. They are allowed to go to the prison courtyard for exercise for two hours per day. Very few can work within the *maisons d'arrêt*. The remaining prisons are mainly *centres de détention* (for those serving long sentences), *maisons centrales* (for dangerous inmates), or *centres pénitentiaires* with different zones (a penitentiary centre can be made up of a detention centre zone and a *maison d'arrêt* zone for instance). Paradoxically, the living conditions are better in these prisons than in *maisons d'arrêt*: no overcrowding, and the opportunity to leave the cell during the day for most inmates.

groups in Russia, and the communal sensibility of Soviet life. Prison personnel in Russia highlight the benefits of communal living, arguing that Russian society still has a 'collective sensibility'; the Head of Education in one prison, interviewed for this chapter, described communal life as a form of preparation for release, pointing out its perceived benefits in preparing inmates for life outside of the colony, and likening the detachment to society as a whole:

> They should get used to living in society. We are preparing them for life in society. They cannot live as if they are on a desert island. They will still be exposed to people, and they should be able to communicate. They must learn to respect those around them. They must learn to understand people, and understand relationships. They must live in the detachment.

These two different modes of prison accommodation provide two different contexts in which to explore the concept of privacy in penal space. Data pertaining to the imprisonment experience in France were obtained through semi-structured interviews conducted in 2008 with 31 male and female adult inmates serving sentences in five different French prisons throughout the country. The data collected for Russia relates to female prisoners within penal institutions across four Russian regions, via over 200 interviews between 2006 and 2010 with prison personnel and incarcerated women and girls, and also outside of these institutions, through interviews with recently released women living in three different cities in European Russia.[3] In both contexts, interviews focussed on life in prison, daily schedules and activities, relationships to family, friends, other inmates and prison personnel. Whereas in France all individual interviews took place without prison personnel present, in Russia all interviews with current prisoners were observed by personnel. A theme which emerged strongly in each context was the experience of penal space, in terms of the lack of privacy experienced by inmates, and this is explored in the following sections of the chapter.

Privacy in Penal Space

In both penal contexts, respondents expressed a desire for privacy, which they described, quite simply, as the ability to be alone. Regardless of the type of accommodation, either the large Russian detachment or the small French cell, prisoners described an all-pervasive lack of privacy. However, in the two different contexts, different aspects of lack of privacy were highlighted as particularly

3 ESRC project 'Women in the Russian Penal System: The role of distance in the theory and practice of imprisonment in late Soviet and post-Soviet Russia' RES-062-23-0026. Data discussed here are also presented in Moran et al (forthcoming) and Pallot and Piacentini with Moran (2012).

troublesome. In France, confinement in small cells was particularly problematic because of the intimacies which are inevitable in this spatial arrangement:

> When there are two of you [in the same cell], it is overcrowding. There is always someone with you. Whenever you eat, wash yourself, go to the toilet, cry, there are always two of you. There is no privacy. It is difficult …
>
> It's very small. If you are with someone else, you have no privacy. You feel stress, anguish. I can't bear it … one of you is shitting and you, you are eating. One is peeing and you are eating. There are many things to endure.

Furthermore, the lack of hygiene in many prisons and the dirtiness of cells played a role in this feeling, as if the fear of contagion – the obsession of architects – is mirrored through the experience of inmates:

> I arrived during the night. I was shocked. [I thought] 'I am going to suffocate. It is disgusting. I will not be able to stay.' It is the fourth dimension, you are in a cage. There is a loss of identity. There is a loss of privacy.
>
> It is dirty. You have no hot water, no kettle, you have nothing when you arrive. You just have a few clothes. It is disgusting, you should see it … Cockroaches everywhere. The sink is yellow. For two weeks, I could not eat. I lost 8 kilos. I didn't go to the toilet. It was hard.

The experience of incarceration is both an experience of confinement in terms of the lack of space – and an experience of dispossession. Prisoners lack a 'safe' place to retreat to – prisoners in France felt that the walls of the cells did not protect them from the carceral environment:

> The noise of keys, of doors, of people shouting, of conflicts, of medical alerts, of people speaking, it is very noisy … There is always noise. It destroys you in the long run. There is always something, a conflict, a suicide attempt; you don't know who is doing it. It is nerve-wracking. I don't want to get used to that. So we put these noises at a distance.

Moreover, the cell itself, even if prisoners are alone in it, is never completely beyond surveillance because of the peephole in the cell door. The peephole can be used by the guards, but is also used by inmates walking in the corridor:

> You can't get used to it … the peephole, the fact … the feeling of being under surveillance, constantly, it is something very, very hard, it is difficult to accept. Sometimes you are surprised by the noise, even if you do nothing special … it makes you twitch … Even when you go to the toilet, you do not feel right. So you put a cloth in front of you, otherwise you don't feel right.

> The peephole, I don't care [about it] personally. They have to have control, to do the surveillance … when it is the guards! Sometimes, it is the others [other inmates] who come back from the courtyard and they look into your cell. That, is rather annoying … we know it is inmates because it is not the time for the guards' patrol.

In Russia, prisoners living in large dormitories also highlighted problems of shared spaces, but rather than pointing specifically to intimacies, they described the psychological pressure of large numbers of people: 'It's difficult, because you can't escape from them [the other women in the detachment]. It's a confined space, and each has their own character, each with their own cockroaches in their head. And sometimes I need to close myself off, to leave them, not to see anyone ...' Although the lack of personal privacy was certainly an issue in Russia, there was a sense from the Russian respondents that the unavoidable 'gaze' of the mass of people in the detachment was different from the very personal observation described in the two-person French cell.

> In essence I am a sociable person, but to talk with people who I just don't find interesting – I don't know, I just don't want to do it. I usually try to hide behind a book, or embroidery ... I try to escape to somewhere. There are one hundred and twenty people [in the dormitory]. You can't even be alone in the toilet! And sometimes you think – God, will there ever be peace? Isn't there anywhere I can be alone?

Although both sets of prisoners described a lack of privacy, it could be argued that in terms of Schwartz' (1972) descriptions of forced exposure, (e.g. communal showers); forced spectatorship (exposure to others' lack of privacy); and violation of collective privacy (imposed and exposed intimacies), the French and the Russian accommodation spaces generated subtly different problems. Although both contexts created the circumstances for all three of these types of lack of privacy, prisoners' comments seem to suggest that in the small French cells, forced exposure, forced spectatorship and exposed intimacies created an extremely acute problem. In the larger Russian dormitories, one might speculate that the mass nature of the observation made the observation somehow less 'personal' and intrusive.

Finding themselves in this context, inmates in both France and Russia sought privacy or solitude in a variety of ways. We discuss these here by arranging them into three categories, each of which represents a varying degree of control on the part of prisoners over their surroundings. We identify the tactics of finding actual spaces in which to be alone, of creating the private in public space, and of retreating into the self even when surrounded by others, and think of these in the following passages as consciously reflective, agentic acts (see also Moran et al forthcoming). In France and in Russia, prisoners deploy and engage with the spaces in which they find themselves in a variety of ways, responding via

the deployment of agentic practice to the specific challenges which they face: we explore these tactics in the following sections.

Finding or Making Private Space

In both penal contexts, prisoners described the tactics they deployed in order to find privacy, or in other words, to be physically alone. In France, the noise and the existence of the peephole cannot be avoided, however the cell was still the space most commonly deployed in this way; a common practice was to decline the opportunity, which French prisoners are offered, to spend time in the prison recreation yard with fellow inmates, so that if a cellmate did choose to go outside, the cell became a private space of sole occupancy for the one staying inside. Prisoners spoke of enjoying the cell 'just for yourself'. One long term prisoner in France described his cell, which he happened to occupy alone, as a refuge against the enforced company of his work as a prison librarian:

> I am stuck at the library, I am forced to be kind, nice to everyone, it's my job, I do it, I am even paid for that, even if it is peanuts, but when I leave the library, I say that's it, I want nothing to do with you. I plunge back into my bubble. At the weekend … I rejoice that I will be in my cell for more than 48 hours. I am lucky that I'm alone, and I would even say that during the weekend, when someone opens the door to bring me food, it pisses me off … I like the evening, when the door is closed. At least, from 6.30pm to 7am, the doors will not be opened. No one will come to piss me off.

The opportunity to decorate the interior of the cell – when permitted– is also an opportunity to re-create a kind of privacy, demarcating a personal space within a shared cell which was not to be violated:

> I have decorated [my cell] with pictures of my children. I feel less alone with them in my cell. I have the impression that there is life. I say hello to them every morning!

> Here [in this prison], we are allowed to decorate [the cells] how we like … Pictures, posters, [it's a way] to have your intimacy … Pictures of your friends, drawings I did here, paintings, pictures of people we like … It is an important decoration for me, because I have my political opinions. My pictures of Sarkozy[4] on the shelves, they were not appreciated by other inmates.

> Normally, we have no the right to decorate the cell, because they think we can make a hole in the wall and hide things behind the posters … But here, the

4 Nicolas Sarkozy was the President of the French Republic at the time of the research for this chapter.

> penitentiary administration is quite ok, as long as you don't put obscene pictures up … I have put up the colours of my home, the colours from the Basque country,[5] flag, posters, pretty girls, but not naked! … I do not deny my country.

These decorations enable prisoners to deploy agency to create a space which is theirs (see also Baer 2005), marked by aspects of their lives beyond incarceration, such as family photographs, or political symbols, and perhaps thereby assist in the creation of a type of privacy within the shared prison cell.

In shared cells, the personal decorations are put up very close to the beds, so as to create a personal space within the shared space. If we understand incarceration in a confined space as a privation of space, there is nonetheless a juxtaposition of space within the cell itself, with a small and very personal space (on and near to the bed) within a larger space which is more difficult to make your own:

> My bed is my living space! I live on my bed, to read and to rest. Since I have back problems, I cannot rest on the stool. Yes, here they have stools, rather than chairs with a backrest. So I stay on my bed. I have everything close at hand, close to the bed. Hence, she [the cellmate] can use the table for herself. My bed is like my lair, really, in the animal connotation of the word. It is the very private part of the cell. It is my spot. If we make a caricature of it, it is my territory, it is the place I do not share. My bed, it is my own space.

In Russia, the nature of shared accommodation spaces, the rules governing these spaces (in terms of restrictions on interior decorations and visible personal effects), and the numbers of people who could potentially be using them at any given time, meant that the kinds of tactics possible in France were not available. For some prisoners in Russia this kind of privacy could only be achieved by gaining control over space in a way which delivered genuine solitude. Such isolation could be achieved by working hard, and being rewarded with one of the few positions of responsibility within the dormitory, which came with certain 'perks'. One former prisoner reported that 'I was a work supervisor, so I had the opportunity to go to the smoking room to smoke during working hours [whereas other prisoners had to go during their breaks from work]. Silence – I'm alone. I enjoyed those five minutes.' Alternatively, for a small number of prisoners, intentionally transgressing prison rules meant a spell of time in the 'cooler' – the isolation cells. '[Other] people shut themselves in the 'cooler', to be alone, they were so sick of it all.' Although the 'cooler' would generally be avoided for its poor living conditions, some former prisoners reported that they chose to offend while in prison; that their desire for solitude could only be satisfied through an instrumental, intentional and strategic

---

5 The 'Basque Country' is the name given to a region that spans the border between France and Spain, comprising the Autonomous Communities of the Basque Country and Navarre in Spain and the Northern Basque Country in France. The politics of Basque nationalism relates to a desire for the self-determination of a single culturally Basque area.

violation of prison rules. '[Committing a violation] was all the same to me. I was so mentally tired, I was like a sponge full of water. Any nonsense, and they quickly lock you up, and it's 10 days, perfect to have time to relax, just to relax mentally.'

Finding Solitude in Public Space

In both Russia and France, prisoners sought privacy through solitude in public space – deploying various tactics to find themselves alone in communal spaces. In Russia, some prisoners reported that they had found quiet spaces beyond the dormitory, to which they retreated in their spare time, after work and at weekends. One inmate described her own tactics, having found a place where, even if she could not guarantee solitude, she could at least escape from the crowds of people in the dormitory; 'Sometimes I need to close myself off, to leave them, not to see anyone ... so I go to the psychological relief room that we have'. Another described another such room in her prison colony:

> We have a room for relaxation. It's in the sanatorium [the prison clinic]. There are psychologists there and you can use the room for relaxation. You can spend an hour and a half there, and that's all. You have to ask the psychologist about how to get into the relaxation room. They've got music, you can pick up CDs, nature sounds, the noise of waves, a storm, some guitar, that kind of thing. We choose something and just sit there.

These spaces were less busy than the dormitory, but they were not places of guaranteed solitude: 'You can sit in the TV room and watch TV – no one will bother you or ask you questions or anything. You can sit somewhere in a corner with a newspaper. Or you can even just sit on your own with a book, no one will bother you'. Although other people were always around, sometimes their mere presence seemed to preclude the kind of privacy that inmates sought; 'There were people everywhere. When you'd go somewhere, there would already be 10 people there, relaxing'.

Privacy in these spaces, therefore, depended on the behaviour and choices of others. None of the spaces mentioned as retreats – such as the psychology room, or the relaxation room, could be booked by individual prisoners, closed off to others who might also wish to share the space, or locked against prison personnel.

Similarly, in France prisoners sought out spaces which, although communal, could be used to generate a semblance of privacy. In the French context a distinction needs to be made between spaces open to very few prisoners – such as the prison garden, or the guards' dining room, where meals are served by a couple of inmates – and spaces open to several dozens of inmates – like the prison workshop – where the atmosphere turns out to be silent, everyone concentrated on the work, and therefore different from the rest of the prison. In both cases prisoners viewed it as a privilege to be there, since it restored their identity as one not fully determined by their incarceration. Prisoners both enjoyed the very small

number of other inmates around, and they appreciated an activity which was not seen as strictly linked to the prison even if they were surrounded by other inmates. The feeling of privacy seemed to arise more commonly in association with the smaller numbers of inmates.

Not surprisingly, the same prison spaces had very contradictory meanings for different inmates: for example, the prison yard was a place of friendship for one, a place of violence for another, a place of dealing (tobacco, drugs) for a third. Privacy is experienced differently in each location: the courtyard can become a place to meet friends and to create a small community; a place of violence; or of drug dealing. However, these spatial practices are connected, since prisoners feel that using the yard in this way, for example to obtain cigarettes or drugs, enables them to make their cells feel more private, since the cells then become spaces to smoke, take drugs, and drink coffee.

> I try to go the courtyard to walk for two hours per day, to bring fresh air to my brain, to breathe, to discipline myself, even if I can't bear the courtyard anymore. There are times when it is very conflictual, with a lot of tension. It is burdensome to be in the middle of quarrels. So there are times when I never go to the courtyard.

> What's the point of going the courtyard? To sit down and wait for an hour and a half? I can't. What's the point? To bring a book? Almost three quarters of those going to the courtyard, they have never read any book, or they mock you for having a book. It is another sphere, another dimension. They, they go because they meet friends. That three quarters, they come from the same [disadvantaged] areas, so they go out because they know one another. But I, I can't. If only there were people to talk to...

> The courtyard is very small … we are not many, and everyone has her spot. Four over there, two or three here … We stay in our spot with our friends. We walk a little bit and come back to sit down and talk … They are used this way, so, well, we will not take her spot.

Each French inmate described having their own time-table, their own use of lived spaces: some went to the recreation area early in the morning to be sure to be alone, or at least to enjoy that space without the company of other prisoners who watched TV all night and slept late in the morning. These spaces become sites of distinction:

> I like the library … there are books, there is contact with the outside, there is contact with the people coming here to give classes, people from outside the prison … not from the Penitentiary Administration …

> At school [within the prison] the teachers are interesting … they are not from your social background … They are nice, the teachers, you speak of other things, they bring you something.

Finding Privacy in the Self

In both Russia and France, prisoners described the ways in which they constructed privacy within public space by retreating into themselves, thus blurring the boundaries between the two.

In France, some prisoners found privacy by retreating into silence. One inmate constantly jogged around the courtyard so as not to have to talk to anyone. Another spent his days in his cell not talking to anyone. He refused an interview, and refused visits from the prison chaplain. This kind of privacy could be argued both to help to support individual identity (the fear of moral contamination) and at the same time to shatter this identity: Le Caisne (2000) has argued that an inmate in France is expected to present many different 'personae' within the prison system. Inmates are expected, on the one hand, to show to the penal administration that they have 'changed' (in order to be granted early release), but on the other to stay the same in the eyes of other inmates, so as not to appear 'weak'. To the prison chaplain the inmate is expected to say that s/he feels closer to God; to the prison social worker that s/he feels ready to rejoin the workforce. S/he must show obedience to the guards when alone, but be rebellious when with other inmates, to 'save face' and to be 'accepted' among the captive community. By the end of the sentence, le Caisne argues, inmates do not know who they truly are. Retreating into the self or finding privacy into the self could be interpreted as a way to protect one's 'true' identity.

In Russia, one public place which facilitated this tactic was the prison factory. Russian prisoners work, and for the majority of female prisoners interviewed for this chapter, this was as a machinist in a prison sewing factory. They worked for seven hours each working day, in a noisy factory packed with individual machines. Although work is undoubtedly hard, for some women it represented a form of solitude, and the insulating noise enabled a retreat from the challenges of communal living into a form of privacy within the self. The workplace within the prison, although not a space where they could genuinely be alone, became somewhere for inmates to at least be alone with their thoughts, even if still surrounded by other people. Some prisoners therefore found solitude in prison work: 'Work helps in that there is time to be alone, alone with your thoughts. Because the dormitory is always in a commotion.' One former prisoner recounted her own advice to fellow inmates, making clear the potential of the sewing factory.

> I always said to my girls 'Girls, learn to sew while you can.' Why? Because there's only one place – at the sewing machine – where you can be alone. And when you know how to sew, no one's pulling at you, you're just sitting there, doing your work, and thinking. So ... the bright ones amongst them, they listened.

Others suggested that pensioners and disabled prisoners who didn't have an obligation to undertake prison work in Russia, sometimes still chose to do so, because of the therapeutic effects of this kind of activity. One prisoner observed;

> Sometimes you want to go somewhere, so basically, for most people it's the factory. There you just work, work, work, and somehow you're distracted from it all ... They [pensioners and disabled prisoners who are not required to work in prison] ask me if they can do things, like sweeping up threads in the factory. They want to work. They help in the *dormitory*. They help in the *dormitory* and they come into the factory. Of course, it's difficult for them without work. They sit in one room, with the television from morning 'til night. It doesn't allow them to relax mentally.

In France, work also seemed to provide a respite of sorts from the intense atmosphere of the shared cell. Although the cell could become a place of privacy when it could be occupied singly, when shared it also became a space of oppression.

> In cells with six or eight inmates, there is a lot of racketeering, a lot of abuse ... It is not easy, is it? Being with seven other people, there is only one TV, not having the same tastes, it is not easy ... Some smoke and stuff. Some sleep, others don't. It is horrible. There are weak people who are forced to buy coffee for others.

Confinement in cells, sometimes for up to 22 hours per day, meant that for many French prisoners any opportunity to escape from the intensively observed spaces of the cell into a more anonymous communal space, was taken. The aim seemed to be to reach places that somehow felt less 'carceral' (such as working in the prison garden, where one existed), being able to associate with other people, having an occupation and earning a small amount of money. Getting out of the cell, expanding the available space, was seen as positive almost regardless of the alternative activity. As one prisoner noted, 'What is very, very annoying here, is to be locked up. Going to the library, working, makes me feel relieved. Because I've managed to break down the confinement, to enlarge my space: from the cell, I've reached the prison.' Another described the importance of work in order to escape the cell, where the feeling of oppression is acute: 'When I was told 'you will be serving the meals'... it enabled me to leave the cell. I was happy, compared with being locked up ... Because here, the primary goal is to be out of your cell. When you're in your cell, you feel very oppressed.'

Conclusion

Considering the concept of privacy within penal space, it is clear that although French and Russian prisoners define privacy in a similar way, in terms of the

ability or opportunity to be alone, the different penal contexts in which they find themselves mean that they deploy different tactics to find or to create privacy.

The empirical material from both contexts would seem to suggest that lack of privacy is indeed a pain of imprisonment, felt by both French and Russian prisoners, in their different types of penal spaces (both in the small cells in France and the large dormitories in Russia) However, the precise ways in which this lack of privacy is perceived differs between the two. In France, where prisoners are confined in small shared cells with peepholes in the doors, the issue seems mainly to be the intensely personal nature of their exposure. In these circumstances there is the forced intimacy of close confinement with one another or a small number of individuals to whom everything is exposed. In Russia, by contrast, exposure seems to be more generic, to everyone at once, in the large dormitory housing many tens of people, but to no one person in particular. Whereas the close confinement and intimate exposure is keenly felt in France, in Russia lack of privacy seems to be expressed more vividly as a result of the sheer number of people with whom communal space is shared.

The tactics deployed to find privacy in each context vary accordingly. In France prisoners seek genuine solitude in their cell, when they can get it, either through remaining behind during exercise hours, or by carefully demarcating 'their' private space within the cell through the use of wall decorations. Alternatively, they seek escape from the cell into the more anonymous, yet still surveilled, spaces of the prison as a whole. In Russia prisoners have fewer, or less routine, opportunities for genuine solitude, and they seem to have become adept at seeking solitude inside the self, within public spaces. In both contexts, then, the 'private' and the 'public' overlap and co-exist with one another, in the myriad of spaces inhabited, negotiated and constructed by prisoners. Although the two contexts differ greatly in their penal architectural forms, the pressures of lack of privacy as a 'functional prerequisite' (Schwartz 1972: 229) and a 'pain' (Sykes 1958) of imprisonment remain, albeit in subtly different forms.

In Russia, an ongoing process of prison reform seeks to phase out the communal dormitory detachment and to replace it with the smaller cells more common in western European contexts, such as France. The head of the Russian prison service, Alexandr Reimer announced in 2010 a process of adaptation of Russian penal establishments, with inmates to be housed in cells accommodating between two and eight people. These reforms are being introduced to improve conditions in correctional facilities, to create additional guarantees for the protection of prisoners' rights, including the right to personal safety, and to bring Russia's prisons up to European standards, rather than in any effort to improve prisoners' privacy. There is a stated intention in the reform process to increase the surveillance of prisoners in Russian institutions through the deployment of more modern technology such as CCTV. Although the forthcoming reforms will fundamentally change the nature of prison accommodation in Russian prisons, the experience of France would suggest that this change in spatial arrangement will alter, rather than remove, the lack of privacy in Russian prisons.

Acknowledgements

Dominique Moran would like to acknowledge the UK Economic and Social Research Council's funding (RES-062-23-0026) for the research leading to her contribution to this chapter. A Visiting Fellowship at the Aleksanteri Institute, University of Helsinki, also provided support to Dominique Moran for the preparation of the chapter. She would like to acknowledge the contribution of Judith Pallot and Laura Piacentini to the thinking which supported her contribution to the chapter.

References

Allen, J. 2006. Ambient power: Berlin's Potsdamer Platz and the seductive logic of public spaces. *Urban Studies* 43(2): 441-455.

Baer, L. 2005. Visual imprints on the prison landscape: A study on the decorations in prison cells. *Tijdschrift voor Economische en Sociale Geografie* 96: 209-17.

Blomley, N. 2005. Flowers in the bathtub: Boundary crossings at the public-private divide. *Geoforum* 36: 281-296.

Demonchy, C. 1998. Architecture et évolution du système pénitentiaire. *Les cahiers de la sécurité intérieure* 31: 79-89.

Demonchy, C. 2000. L'institution mal dans ses murs, in *La prison en changement*, edited by C. Veil and D. Lhuilier. Ramonville-Saint-Agne: Érès, 159-184.

Demonchy, C. 2004. L'architecture des prisons modèles françaises, in *Gouverner, enfermer. La prison, un modèle indépassable?*, edited by P. Artières and P. Lascoumes. Paris: Presses de Sciences Po, 269-293.

Dirsuweit, T. 1999. Carceral spaces in South Africa: A case study of institutional power, sexuality and transgression in a women's prison. *Geoforum* 30: 71-83.

Drummond, L.B.W. 2000. Street scenes: Practices of public and private space in urban Vietnam. *Urban Studies* 37(12): 2377-2391.

Fenton, L. 2005. Citizenship in private space. *Space and Culture* 8(2): 180-192.

Foucault, M. 1979. *Discipline and Punish*. New York: Vintage.

Foucault, M. 1991. Governmentality, in *The Foucault Effect: Studies in Governmentality*, edited by G. Burchell, C. Gordon and P. Miller. Chicago: University of Chicago Press, *87-104*.

French Ministry of Justice, 1990. *Programme pour la construction du centre pénitentiaire de Ducos (Martinique)*. Paris: Ministère de la Justice.

Grant, E. and P. Memmott 2007. The case for single cells and alternative ways of viewing custodial accommodation for Australian Aboriginal peoples. *Flinders Journal of Law Reform* 10: 631-647.

Jackson, P. 1998. Domesticating the street: The contested spaces of the high street and the mall, in *Images of the Street: Planning, Identity and Control in Public Space*, edited by N.R. Fyfe. London: Routledge, 176-191.

King, R.D. 1994. Russian prisons after perestroika – End of the Gulag? *British Journal of Criminology* 34: 62-82.

Kumar, K. and E. Makarova 2008. The portable home: The domestication of public space. *Sociological Theory* 26(4): 324-343.

Le Caisne, E. 2000. *Prison. Une ethnologue en central*. Paris: Odile Jacob.

Milhaud, O. 2009. Séparer et punir. Les prisons françaises: mise à distance et punition par l'espace. Unpublished PhD thesis, University of Bordeaux, Bordeaux.

Mitchell, D. 2003. *The Right to the City: Social Justice and the Fight for Public Space*. New York: Guilford Press.

Moran, D., J. Pallot and L. Piacentini (forthcoming) Privacy in Penal Space: Women's Imprisonment in Russia *Geoforum*

Oleinik, A.N. 2003. *Organized Crime, Prison, and Post-Soviet Societies*. Aldershot: Ashgate.

Pallot, J, and L. Piacentini with D, Moran 2012. *Gender, Geography and Punishment: The Experience of Women in Carceral Russia* Oxford University Press, Oxford

Salle, G. 2011. La maladie, le vice, la rebellion. Trois figures de la contagion carcérale. *Tracés* 21: 61-76.

Schaeffer, M.A., A. Baum, P.B. Paulus and G.G. Gaes 1988. Architecturally mediated effects of social density in prison. *Environment and Behaviour* 20: 3-20.

Schwartz, B. 1972. Deprivation of privacy as a 'functional prerequisite': The case of the Prison. The *Journal of Criminal Law, Criminology and Police Science* 63(2): 229-239.

Sharkey, L. 2010. Does overcrowding in prisons exacerbate the risk of suicide among women prisoners? *The Howard Journal* 49(2): 111-124.

Sibley, D. and B. Van Hoven 2009. The contamination of personal space: Boundary construction in a prison environment. *Area* 41(2): 198-206.

Simon, J. 2005. The return of panopticism: Supervision, subjection and the new surveillance. *Surveillance and Society* 3: 1-20.

Staeheli, L.A. 2010. Political geography: democracy and the disorderly public. *Progress in Human Geography* 34(1): 67-78.

Sykes, G.M. 1958. *The Society of Captives: A Study of a Maximum Security Prison*. Princeton: Princeton University Press.

Tyndall, A. 2010. 'It's a public, I reckon': Publicness and a suburban shopping mall in Sydney's southwest. *Geographical Research* 48(2): 123-136.

Vaz, P. and F. Bruno 2003. Types of self-surveillance: From abnormality to individuals 'at risk'. *Surveillance and Society* 3: 272-91.

Chapter 13
Resisting 'Bare Life': Prisoners' Agency in the *New Prison Culture* Era in Colombia

Julie de Dardel

Introduction

From the beginning of the twenty-first century, the Colombian prison system has gone through significant alteration in both scale and in nature. As the prison population tripled, the embedded prison culture underwent a dramatic change as a result of the *New Prison Culture* reforms – a 'modernization' policy inspired by the US Federal Bureau of Prisons. The implementation of the US model did not just represent macro-changes in the Colombian prison infrastructure and rules: it has had very concrete effects on the prisoners' experience of incarceration and the manner in which they deploy their agency in this much more restrictive setting.

In this chapter, I analyse how the principles of the New Prison Culture collided with the embedded practices and representations of the old Colombian system, since it marked the start of an unprecedented management of spaces and bodies. The material and symbolic dispossession lived by the inmates in a situation of severe and arbitrary control echoes what philosopher Giorgio Agamben defines as 'bare life', a life deprived of political status and subjected to a permanent 'state of exception' (Agamben 1998). I draw on his theoretical hypothesis – rarely used in prison studies so far – to associate his conception of the 'camp' as a contemporary paradigm of biopower exemplified in the functioning of the New Prison Culture facilities in Colombia.

I claim however that Agamben's gloomy and univocal vision of the camp is challenged by ethnographic data on the Colombian prison system which reveals the persistence of an active resistance to the sovereign power exercised within carceral space. Even in the very restrictive setting of the new regime, prisoners manifest agency by subversively using materials, times and spaces to create an 'underlife' in the total institution (Goffman 1961), particularly by developing tactics of the body – or more specifically *tactics to clothe the body* – to oppose *bare life*. Goffman's concept of 'secondary adjustments' brings here a fundamental theoretical support to approach daily resistance inside the prison.

I also argue that there is some leeway between the inmates and the prison authorities which gives space for manoeuvre in which the inmates can negotiate part of the rules in the total institution. Thus, I show that the New Prison Culture reform partly failed to eradicate the alleged vices of *criolla* culture, i.e. the

universe of representations and practices that underlies the old local prison system. This situation ended up creating a form of permeation between two distinctive universes of customs and meaning.

This chapter draws on original empirical material from the under-researched penal context of Latin America. It aims to contribute to the renewal of ethnographic perspectives in prison studies, after a puzzling eclipse of empirical research since the 1980s, at the very time the prison sector underwent an historical boom worldwide (Wacquant 2002, Rhodes 2001). In recent years, geographers have revived interest in prisons and ethnographic methods, by turning their attention to the inmates' world and giving significance to prisoners' agency (Dirsuweit 2005, Baer 2005, Pallot 2007; Sibley and Van Hoven 2008). However, in the last two decades, prison studies have been mostly explored through a macrostructural perspective: an important body of knowledge has been gathered on the expansion of the prison complex, which has been described by critical scholars as a key institution of the contemporary era, an important instrument of post-welfare state-building, a powerful regulatory means to manage the most precarious and segregated classes of society, as well as a fundamental device in the deployment of new control, enforcement and surveillance technologies (see for example Christie 1994, Garland 2001, Peck 2003, Peck and Theodore 2008, Gilmore 2007, Pager 2007, Wacquant 1999, 2009). The renewal of the ethnographic approach in this research field can complete and enhance this viewpoint and play a significant role in the critical understanding of the increasingly central place of the prison in the broader social, economic and political context of late capitalism. The rich comparative account offered by the Colombian case can certainly contribute to that aim.

The evidence employed here draws on extensive fieldwork within the Colombian prison system that took place between October 2009 and February 2011. Methods included direct observation in prisons and in-depth interviews with prisoners, former prisoners, family members, prison officers, authorities, architects and Human Rights defenders. The chapter presents three accounts selected out of a series of 18 interviews carried out with serving prisoners in different types of correctional facilities, which serve to illustrate prisoners' agency in the very restrictive setting of the New Prison Culture. Semi-directive interviews of one and a half to two hours in length were conducted in a confidential setting (in the absence of prison officers or other prisoners); for security reasons, tape recorders were not allowed into the prison and interviews were not transcribed verbatim, but rather were written up from detailed handwritten notes taken during the meeting with the prisoner. Even though the cause of imprisonment is not directly relevant to the issue at hand, it should be pointed out that these three people are 'political prisoners' who were convicted for illegal activities as members of one of the Colombian guerrilla organizations, and are serving long sentence (over 20 years).[1]

1 In order to protect sources, names and other identifying data have been altered.

Access to the prisons was gained thanks to a Colombian Human Rights NGO whose representatives accepted that I join their humanitarian visits to the prisons, so I could conduct fieldwork alongside their activities. The national prison authority, INPEC (*Instituto Nacional Penitenciario y Carcelario*), was informed about the research and also participated in several interviews. The relative openness and flexibility of the INPEC in relation to this research is undoubtedly inherited from the *criolla* prison culture, which tempers the increasing strictness of the new high security regime.

In what follows, I first put the New Prison Culture in the context of 'Plan Colombia', the US political and military intervention in this country, and highlight the tensions between the old and the new prison system. I then analyse to what extent the American model implemented in Colombia presents similarities with Agamben's notion of the camp, before challenging his totalizing and dark perspective by considering three examples of active agency and resistance to the bare life within Colombia's prisons.

The Cultural Embeddedness of the Colombian Prison System

Over the last 20 years, Colombia has undergone a dramatic expansion of its prison system. As with many South American countries, economic neoliberalization in the 1990s, labour deregulation and privatization of public services, have coincided with an unprecedented tightening of the screws in relation to criminal law and repressive practices (Wacquant 1999). As Colombia became the most unequal country in South America,[2] the government decided to attend to criminality, as well as social and political unrest, with an iron fist. As a result, between 1994 and 2011, the Colombian prison population tripled to reach 100,000 inmates and in the last decade the prison budget increased fourfold.

At the beginning of the 2000s, Colombia and the United States joined forces to undertake reform of the Colombian prison system: this new policy was termed the New Prison Culture (*Nueva Cultura Penitenciaria*). The agreement (*Programa de mejoramiento del sistema penitenciario colombiano* 2000/2001) was carried out as a part of Plan Colombia, the $7.5 billion anti-narcotics and anti-insurgency US programme for this country. It stipulates the technical input of the US Federal Bureau of Prisons (BOP) in the design, construction, equipment and internal rules of new high security correctional facilities in Colombia. Initially, the prisons built under this agreement were modelled on the U.S. federal facilities in Coleman, Florida (Wilkey and Rivera 2002). Between 2001 and 2004, six large new Colombian prisons were constructed during the project's first stage, all characterized by their isolated locations, an absolute priority given to security, draconian rules and harsh

2 Colombia's national Gini coefficient (measuring inequality of wealth) for 2009 was the highest in South America (ECLAC 2010).

treatment of inmates. The BOP ended its mission in Colombia in 2004, having recommended 10 more new correctional mega-complexes be built by 2010.

From its inception, the New Prison Culture promised to confront the old prison system by bringing an end to the laxity, loss of control and disorder that characterized it. At the turn of the century (i.e. when the BOP came on the scene), the Colombian prison sector was going through an unprecedented crisis. The penal system was suffering from dilapidated infrastructure, extreme levels of corruption, delinquency, poverty and violence, absurd privileges for crime bosses, as well as an endemic lack of control by the State. In practice, prisons were jointly managed by both powerful inmates and the authorities. Furthermore, as a direct consequence of the punitive turn of the 1990s, Colombian prisons began to reach unthinkable levels of overcrowding. The situation was such that in 1998 the Constitutional Court of Colombia declared the whole prison system to be unconstitutional (Sentence T153 1998).

However, this view of the old prison system would be incomplete if not balanced by consideration of some culturally-rooted practices that have greatly contributed to raising the quality of life behind bars. Indeed, penal institutions are embedded in cultural and institutional traditions that necessarily shape the way an outside model is adapted in the new local setting (Melossi 2004). In Colombia, a set of customs and practices regionally shaped a particular form of life behind bars and formed what can be defined as the *'criolla'* culture, a term broadly used by the prisoners and the prison officials themselves to describe the traditional penitentiary subculture. The opposite of the isolation pattern of the New Prison Culture, the *criolla* culture was historically rooted in practices of integration into the Colombian society, such as: (1) a prison system open to the outside world and allowing regular and close contact with family and society; (2) the location of the prisons, generally integrated into easily accessible urban settings; (3) a strong communal life style letting the prisoners collectively self-manage almost all aspects of their daily lives; and (4) some degree of freedom in relation to the prison authorities, as well as a substantial margin of dialogue and negotiation with the latter. The *criolla* prison culture afforded – and still affords in the vast majority of the facilities in the country where the New Prison Culture has yet to be implemented – some clear benefits which help preserve social relations and moderate the depersonalizing effects and 'mortification of the self' imposed by the total institution (Goffman 1961). Evidently, the *criolla* prison model is itself embedded in the more general framework of 'Latin American culture' and, particularly, in the Colombian way of life. Some cultural patterns – like the centrality of the institution of the family; the emphasis on relations of seduction and romance; the reign of informality (implying both a loose relation to the rules and a remarkable resourcefulness in everyday life); as well as the impacts of the armed conflict and the related strength of the social movement – definitely mark the prison subculture of that country. The courtyard is the space where the collective and self-organized life of the prisoners takes place, with very little presence of the prison officers (Figure 13.1).

Figure 13.1 Daily scene in a *criolla* prison, Columbia
Source: Comité de Solidaridad con los Presos Politicos (FCSPP), Colombia.

Undoubtedly, the introduction of the US prison model represented a dramatic change compared to the old national prison culture. The new system immediately collided with the profound informality and the social integration of the *criolla* prison system. The new scheme imposed an unprecedented blueprint of close control and extremely severe restrictions, which implied a whole new management of spaces and bodies. New Prison Culture uses techniques to separate, isolate and cut communication channels. The location in remote areas; high security architectural design (large facilities divided in small units with high concrete walls); drastic limitations on family visits; virtual absence of rehabilitation programs; and brutal use of force according to the 'New Correctional Techniques', largely contributed to a rising feeling of disconnection and desocialization among the inmates. Moreover, the broad list of deprivations – among them the compulsory wearing of uniforms and the shaving of head and facial hair ; prohibitions on using a watch or a mirror, listening to music, or having more than three books and three photographs of family members – represents an attempt to strip the prison population of its human attributes, or at least its attributes as citizens. In contrast, despite chaotic and dangerous conditions, the *criolla* prisons did not aim to exclude the prisoners and subject them to a totalitarian form of control. To that extent, the violence of the 'Americanized' prisons in Colombia is of a different nature, tending to reduce

the prisoner to a biological survival; in this, they strongly echo the bare life of the 'camp', a notion defined by Italian philosopher Giorgio Agamben in his highly influential work *Homo Sacer* (1998).

A Biopolitical Space Challenged by Acts of Resistance

In recent years, the original studies of Giorgio Agamben on sovereign power and biopolitics in the modern era have had a major impact on critical migration and refugee studies (Walters 2008), but have drawn less attention from scholars engaging with penitentiary spaces. This is surprising because Agamben's work would seem highly relevant in this context.

Agamben (1998) claims that sovereign power in contemporary times increasingly coincides with the exercise of biopolitics that tend to reduce citizens to a form of 'bare life' and maintain them in a generalized 'state of exception'. His demonstration rests on a genealogy of sovereign power, which he traces back to the Roman Empire. Bare life is originally the life *homo sacer*, a banned being who, according to the Roman law, is not eligible for ritual sacrifice, but can be killed by anyone with impunity (Agamben 1998: 12). Hence, bare life is defined by this double exclusion from political and sacred space, i.e. a life reduced to biological existence stripped of every political status and immersed in a 'state of exception' where the rule of law is permanently suspended (Agamben 1998: 96). The very existence of this in-between space 'at once excluding bare life from and capturing it within the political order' (Agamben 1998: 12) represents the 'hidden foundation' of sovereign power and its hold on life until today. Indeed, spaces of exception where bare life prevails have been constantly reproduced since the time of *homo sacer* and represent the 'matrix and *nomos* of political space in which we are still living' (1998: 99). Agamben uses the notion of the 'camp' to conceptualize the multiple concealed forms in which this space of exception has continued to be materialized: 'The camp – as the pure, absolute, and impassable biopolitical space (insofar as it is founded solely on the state of exception) – will appear as the hidden paradigm of the political space of modernity, whose metamorphoses and disguises we will have to learn to recognize' (Agamben 1998: 73). Current antiterrorist laws, Guantanamo prison or detention centres for smuggled immigrants – among numerous examples – are cases where permanent states of exception and conditions of bare life are materialized (Agamben 2005).

Though considered one of the most acute analysts of post-9/11 (Bigo 2007), Agamben's radical viewpoint on the latent slide into fascism of the democratic regime – or rather the way he considers that totalitarianism and democracy 'cross over into one another since they ultimately rest on the same foundation' (Heins 2005) – raised opposition among scholars. Criticisms of Agamben can be summarized under three headings: his excessively nightmarish and monolithic vision of the modern spaces of bare life, embodied in the camp (Levy 2010); his lack of proportion when he includes, in a single 'truth of political modernity'

(Mesnard 2004), varieties of spaces such as refugee camps, Auschwitz and gated communities (Levy 2010); and his failure to recognize the complexity of survival inside the camp, in particular acts of resistance and capacities to keep on being human (Bigo 2007, Walters 2008).

From my perspective, Agamben's conceptualization of the camp remains an enlightening approach to the new High Security prisons in Colombia. Indeed, the New Prison Culture has created a biopolitical space where sovereignty overpowers life itself and where the subjects are stripped bare both materially and symbolically. Prisons inspired by the US model in Colombia have attempted to dispossess inmates of their sense of self by the destruction of personal privacy and the permanent exposure of the body. Geographical, social and affective isolation of prisoners have reinforced a feeling of banishment. Furthermore, prisons of the New Culture meet conditions of the 'camp' insofar as they represent an in-between space, at once included within the political order and excluded from it, in which reigns a permanent 'state of exception'. The suspension of the rule of law is notorious, accompanied by arbitrary internal rules; the brutality of special commandos trained by the US instructors; and the impunity that systematically covers acts of bad treatment or torture of prisoners.

However, in agreement with Agamben's critics, I concur that this bleak vision of the camp should be counterbalanced by considering more closely the practices of resistance that form the underlife of the prison and constantly defy the control exerted over the subjects. Therefore, putting Giorgio Agamben into conversation with Erving Goffman can offer an interesting viewpoint that takes into account both the attempt of the carceral space to impose bare life and the agency deployed by the inmates in that setting. In *Asylums* (1961), Goffman places a strong emphasis on the inmates' faculty for self-preservation and their capacity for putting some distance between themselves and the expectations of the 'total institution'. Hence, he develops a series of concepts that allow us to grasp the total institution's underlife, in which 'secondary adjustments' – a set of improvised means, resourceful use of objects, times and places – are used by inmates to clandestinely preserve a certain margin of freedom (Goffman 1961: 189). His theory explicitly includes the idea of subversion of the system, since he sees the total institution and its clandestine insubordinate underlife as intrinsically bound together.

This theoretical approach to the prison as a space of both subjection *and* subversion underpins the analysis of the empirical findings in the next part of this chapter. This ethnography of the Colombian prison system backs up the idea that survival inside the camp is more complex than Agamben claims, and shows that inmates' agency does defy sovereign power inside the space of exception. In addition, given that the living body is the target inside the biopolitical space of the 'camp', it is not surprising that many of the prisoners' acts of resistance precisely use the body to resist bare life behind walls. The use of the body as a means of opposition in prison has been experienced in the past in other historical contexts. One of the most outstanding cases is undoubtedly the struggle of political prisoners

in Northern Ireland who used the body as a 'weapon artefact' and converted their political agency into an 'embodied force' (Feldman 1991: 2).[3] In a similar way, the Colombian prisoners' bodies have become the core of their resistance, as I argue in what follows.

Refusing 'Naked Life': Prisoner' Tactics to '*Clothe the Body*'

In March 2010, I interviewed Milena, a 40-year-old female prisoner in the High Security prison of Valledupar, the harshest facility in Colombia, located in an isolated area of an arid region on the northern coast.[4] When I asked her what she would change if she could redesign the prison, she thought for a long while and then responded: 'I just can't think of anything. Here, in the prison, they succeeded in locking up our minds. They killed our imagination'. Nevertheless, despite Milena's affirmation, her narrative and those of many other prisoners are far from showing such lack of imagination: even inside the dehumanized environment of the Americanized prisons, the inmates keep on inventing hundreds of 'tricks' and daily micro-transgressions, which taken together, represent a kind of low-intensity everyday resistance to sovereign power in carceral space. These marks of opposition are similar to the cases of detention camps for unauthorized migrants at European borders: inhabitants of the camp are not passive, helpless beings – as Agamben may hint – but also active agents defying sovereign power (Walters 2008). In this section I draw on interviews with Milena, Tania and Fernando, three individuals whose experiences eloquently demonstrate the complexity of the underlife in contemporary Colombian prisons, despite the conditions of severe deprivation and the arbitrariness of the New Prison Culture.

When Preserving Femininity Becomes an Act of Resistance

Like the women in prisons in South Africa (Dirsuweit 2005) or Russia (Moran and Milhaud, this volume), women prisoners in Colombia are generally forced to conform to a discriminating gender model of ideal 'womanhood', especially in the kind of rehabilitative, educational and recreational activities to which they have access. However, the women's block of Valledupar High Security Prison was an exception to this general pattern, as the punitive goal of seeking the mortification of the inmate's self replaced any attempt at 're-education'. In such circumstances,

3 Techniques of resistance of the prisoners in Northern Ireland included the refusal to wear uniform, the staining of the cells with bodily waste and a hunger strike causing the death of 10 men in 1981 (Feldman 1991: 221).

4 The women's block at Valledupar High Security Prison was closed shortly after my visit, following a national scandal over the inhumane conditions in which the female inmates were kept. Despite repeated protests by inmates and an international campaign to close this prison, 1600 male inmates are still held there in similar conditions.

rather than aiming to constrict women inmates to a feminine model as is the case in other Colombian correctional institutions, the prison rules were intended to attack the very basis of 'womanhood' as part of their identity. In Valledupar, women were subjected to a tight biopolitical control of their body, being strictly forbidden to make any kind of outward expression of their femininity, such as wearing jewellery, make-up, perfume or female attire, and compelled to wear the same masculine uniforms as the men. A significant part of the underlife of the female population was thus dedicated to clandestinely maintaining their identity as women, which paradoxically in this particular setting became an act of subversion rather than conformity. The tactics Milena told me the women use to circumvent prison rules and maintain their femininity are striking examples of active agency inside the 'camp':

> The hardest thing for us has been the ban on wearing earrings. We couldn't stand it any longer, so we started to gather feathers that fell from the sky into the yard. And by combining them with bristles from the brooms, we make earrings, which we use clandestinely. Makeup enters illegally, but it is very expensive. We have to use normal pencils to put it on. As for clothes, the women started to go down from the cell to the yard each morning bringing their own 'civilian' clothes under their arms. Once the guard finishes roll call, we just change into our own clothes until 4 pm, when we have to put the uniform back on again to return to the cell. The warders ended up accepting this, although they arbitrarily change their mind some days and write our names down in the disciplinary register under the pretext that we are looking indecent.

Chronic paucity of access to basic services – such as water and health care – is another serious issue in Valledupar that strongly echoes the circumstances of 'naked life' and biological survival Agamben talks about. However, even in these conditions of deprivation, women inmates struggle to improve their living conditions and be able to keep themselves nice and clean:

> During the day, they only open the water supply twice, for 5 minutes each time. But we noticed that we could get more water by sucking it up with our mouths from a few broken pipes jutting out from the bathroom wall. That way, we can wash during the day, and fill up small tanks of water that we bring back to our cells to use at night.

Maintaining Intimacy and Seduction

Another significant illustration of resistance implying subversive use of space and the body occurs in the correctional complex of Jamundi, located 24 kilometres from the city of Cali (Figure 13.2). With 4200 beds (1000 of them for women), it is the biggest of what the Colombian government termed the 'latest generation prisons', based on the US federal complexes. Jamundí prison includes several

prisons in one, comprising all security stages, as well as pre-trial and sentenced prisoners. The presence of the opposite sex in the same detention facility is far from insignificant for male and female prisoners, and the attraction of this separate but palpable presence generates very interesting developments in the underlife of the institution. Notably, the possibility of a blurred but still direct visual contact between the men confined in the highest floors of the remand towers and, down below, the women locked up in the distant female blocks, has led to the setting up of a unusual Goffmanian 'free place', escaping the total institution control and loaded with a manifest erotic content.

Figure 13.2 Prison complex of Jamundi, Columbia
Source: Periodico *El País*, Ernesto Guzman Junior, Cali, Colombia.

I interviewed Tania in November 2010 whilst visiting Jamundi prison. Although we were meeting in the visiting yard set apart from the blocks, our conversation was interrupted by the high level of noise around us. I immediately asked my respondent why women were yelling so loudly and if there was a particular problem. 'Oh that?' she said, 'It is just the *chatroom*'. Tania then patiently

explained to me how this flirting communication actually worked. It was similar to a chatroom on a dating website, between inmates who have never met but can catch a glimpse of each other behind their respective bars – but in this case the language involves the whole body. I asked her to write down the rules of the courtly exchanges, and a few weeks later I received a text about the 'prison chatroom'. Here are some extracts:

> Every day, a group of women mainly under the age of 25 nimbly climb up the large windows and sit on the narrow lintels, sticking their arms and legs out through the bars, and get themselves comfortable to be able to get through the day. Once they are installed, they flutter their towel to the men 400 meters away who, like them, also have a towel ready to start the 'prison chat': an interactive communication where the keyboard is replaced by a towel skilfully coiled in order to precisely form alphabetical figures with the support of their arms and legs.
>
> With great patience the most skilled women teach the newcomers to make each letter, explaining that orthographic spelling shouldn't be used, and that if you make a mistake you should shake the towel as if you were cleaning a blackboard. To show anger, the towel is shaken several times, as if you were beating something. Laughter is the easiest thing to express: you just need to write 'lol' [Laugh Out Loud]. To send kisses, you bring the towel to the mouth and then you make the gesture of blowing on it.
>
> At first, women start spelling as if they were at preschool. But with greater practice, they manage to write very adroitly. Obviously, each woman has her own style and gives a particular rhythm to her handwriting. The shape in which the towel is held, the width of the letter and the speed are part of a person's stroke. Those details don't go unnoticed to a young watchful and eager eye, since they are indispensable to recognize each other and detect possible imposters.

Tania's account, as well as Milena's, not only demonstrate the broad employment of Goffman's 'secondary adjustments' – through the subversive use of materials and places – in the present everyday life of the total institution; more interestingly, they evidence how the prisoners still deploy agency to refuse bare life, even in the extremely restrictive setting of 'perpetual state of exception' prevailing in the New Prison Culture. On the one hand, prisoners' stories reveal how tactics of the body, or more specifically *tactics to clothe the body*, are used to oppose the nakedness – in both metaphorical and in literal sense – imposed by the prison. Indeed, the two female prisoners' accounts highlight how they confront the total institution by resourcefully using devices to maintain their femininity, hygiene, and dignity, in a space that aims to dispossess them of those constituent elements of their identity. Tania's account reveals another facet of the use of the body as a

means of resistance: they report how prisoners struggle to preserve their intimacy, sexuality, and seduction with partners of the opposite sex.[5]

On the other hand, the ethnography of the Colombian prison system reveals the importance of considering the finer scale of the interpersonal embodied relationships to understand the daily reality of the institution beyond official discourses (Mountz 2004). Prisoners' accounts demonstrate that there is leeway in the relation between the community of the prisoners and the prison authorities that allows the creation of a power struggle between the inmates and the authorities. Thus, the role that the prison officers play inside the 'camp' is not univocal and, to a certain extent, can even enable the prisoners' resistance. The interviews show that the warders often turn a blind eye to the inmates' bending of the rules and that repeatedly banned behaviours often become tolerated. This room for manoeuvre is in turn used by the inmates to counter *bare life* in the carceral space. Additionally, more confrontational ways of resistance can be successful too, as it is explained in what follows.

Leeway for Negotiations: A Legacy of the *Criolla* Prison Subculture

Acts of protests are far from unusual in Colombian prisons, since there is a strong tradition of social movement within this population, fuelled by the presence of numerous and organized political prisoners from the left-wing guerrillas. Resistance goes beyond 'secondary adjustments', or rather combines with them, to establish a balance of power likely to change the rules of the camp. It was expected by US and Colombian authorities that the drastic conditions of the New Prison Culture would be successful in eradicating such practices, this being one of its declared goals: but it has not, as Fernando's account demonstrates.

Fernando is a 36 year old prisoner who has spent the last seven years in Combita Prison, another US model facility located in the mountainous and cold Andean region of Boyaca. As he explained to me when I met him in May 2010, although Combita continues to be one of the toughest prisons in Colombia, the rigid US rules have relaxed considerably since the prison first opened in 2002, under pressure from both the 'low intensity' daily resistance and the open collective struggles of the prisoners. His account brings to light the nature of the leeway for negotiation that exists between the inmates and the prison authorities.

Like other inmates at Combita, Fernando is not forced to remain in the courtyard during the day, since he can go back to his cell when he wants to. Indeed, he even has his own key, something unthinkable in the other US model prisons in Colombia.

5 Because of the limited extent of this chapter, I don't discuss homosexual relationships within the female inmates' population, although they are widespread in Colombian prisons, and represent another way of refusing bare life.

> A few years ago, a prisoner took apart the lock of his cell door. He succeeded in making a mould and made his own key. From that time on all the prisoners started to make their own keys. We make them by melting little pieces of plastic, taken from the likes of tables and chairs. At first, the guards opposed this, but we insisted and they ended up letting us have our own key, if we need to have a moment of privacy.

Fernando told me that the requirement to wear the uniform was also overturned thanks to the same kind of *fait accompli* actions: the inmates just stopped wearing it, and after a while, the prison management stopped compelling them to do so. Nevertheless, in Combita, most of the small gains made by the inmates were won through open and collective resistance and prisoners' strikes. In such a manner, they improved their living conditions and got access to various banned items and activities, such as arts and crafts in the yard; having a radio and a collective coffee machine; using chairs and tables in the courtyard and in the visitor's area so they did not have to sit on the ground the whole day. One significant prisoners' strike took place in 2004 precisely about an issue involving the body and sexual intimacy: the strikers demanded improvements to the conditions for conjugal visits. Indeed, this is probably the most important institution of the *criolla* Colombian prison system (the conjugal visit still takes place each week in the old facilities in Colombia). Fernando told me:

> At that time, we had a conjugal visit every 45 days, and though we officially had an hour, the warders began to annoy us by knocking on the door after 25 minutes to throw us out. Those visits were a very traumatic experience for both the prisoners and their lovers. At one point, we decided to go on strike. The whole prison took part. For two weeks we refused prison issue food, did not take part in roll call and refused to cooperate with lock-up. We slept outside, in the courtyard, all of us together. The warders raided us to take away all the blankets. We were freezing at night, so we slept huddled together and we had to pace round at times because of the cold. We faced tough repression: they threw in teargas, they beat us, but we stuck it out and in the end they negotiated. Each block sent a representative to talk to the prison director. Since then, there is a monthly hour and a half intimate visit.

Milena, Tania and Fernando's accounts reveal different facets of the refusal of 'bare life' within the total institution and 'tactics to clothe the body' the prison population oppose to nakedness, both literally – i.e. by using make up, wearing their own outfits instead of the uniform, maintaining personal hygiene – and symbolically – by struggling against material and emotional dispossession, preserving their intimacy and sexuality, asserting themselves as political beings with specific needs. In these tactics, there is a space for collaboration between the inmates and the rulers, since many of the prisoners' initiatives are tolerated and a lot of their claims are eventually approved. Those acts of resistance, as well as the

leeway of negotiation described above, are undoubtedly rooted in the Colombian *criolla* prison culture. Despite the unshakeable resolution to wipe out the 'bad habits' of old prisons while imposing the New Prison Culture rules, over the years, many embedded practices of the former system have infiltrated the high security 'Americanized' facilities, creating a hybrid system where two contradictory carceral subcultures interpenetrate.

Conclusion

The ethnography of the Colombian prison system is a compelling case for prison studies that contributes to the critical analysis of the neoliberal model of prison management and its growing influence at an international level. Furthermore, this example brings to light the significance of local culture in the way a model is implemented in a new setting. In Colombia, the embedded practices and representations resulting from the *criolla* culture characterizing the old prison system have limited the impact of isolation and desocialization inherent in the prison model implemented by the US Federal Bureau of Prisons.

Thus, if Agamben's theorization of the camp provides an illuminating framework to analyse the display of sovereign power in the expanding spaces of the state of exception – going from the confinement zones for undocumented migrants at the doors of the 'first world' to the maximum security prisons of the post-September 11th era – it remains blind to other fundamental aspects of the functioning of the 'camp' such as the role of cultural factors, which are, in the Colombian case, intrinsically linked to the development of tactics of resistance by the inmates' community. The accounts of Milena, Tania and Fernando counterbalance Agamben's monolithic vision by disclosing the multifarious ways in which they oppose the close control imposed by the New Prison Culture. Those acts predominantly lean on tactics of the body that reveal the permanence of their autonomy and, to a certain extent, of their rebellion. Therefore, Goffman's enlightening approach to total institutions, as it emphasizes the multifaceted mechanisms by which the inmates constantly subvert the system and express agency, offers a complementary viewpoint to Agamben's concept of the camp. The association of Agamben and Goffman's perspectives is particularly relevant to grasp the complexity of carceral space.

Ethnographic research methods employed in this study undeniably uncover the contradictions and the power relations that underpin carceral space. The latter can no longer be seen as a place of totalitarian control neutralizing subjects and reducing them to the biological existence of *homo sacer*. Empirical material shows how prisoners deploy agency to preserve their identity, but also unveils how prison officers' behaviours can ultimately facilitate the refusal and resistance of the inmates.

Despite record levels of incarceration in many countries of the world, little is known of the life behind bars. The study of the Colombian prison system, although

based on a very specific context, suggests that conducting empirical research in the heart of the prisons is a fundamental means to understand the reality of imprisonment in the contemporary era. This research aims to contribute to the building of a comparative ethnography of the prison space that is not disconnected from the experience of people living behind those walls, and gives central attention to their faculty to deploy agency in this restrictive setting. Hopefully their perspective will continue to be taken into account in further work on carceral spaces.

References

Agamben, G. 1998. *Homo Sacer: Sovereign Power and Bare Life*. Stanford, CA: Stanford University Press.

Agamben, G. 2005. *State of Exception*. Chicago: University of Chicago.

Baer, L. 2005. Visual imprints on the prison landscape: A study in the decorations in prison cells. *Tijdschrift voor Economische en sociale Geografie* 96(2): 209-217.

Bigo, D. 2007. Exception et ban: A propos de l'Etat d'exception. *Erytheis* 2: 115-145.

Christie, N. 1994. *Crime Control as Industry: Towards Gulags, Western Style*. London: Routledge.

Dirsuweit, T. 2005. Bodies, state discipline, and the performance of gender in a South African women's prison, in *A Companion to Feminist Geography*, edited by L. Nelson and J. Saeger. Oxford: Blackwell, 350-362.

ECLAC, 2010. *Statistical Yearbook for Latin America and the Caribbean*. Santiago de Chile: United Nations Publication.

Feldman, A. 1991. *Formations of Violence: The Narrative of the Body and Political Terror in Northern Ireland*. Chicago and London: University of Chicago Press.

Garland, D. 2001. *Mass Imprisonment: Social Causes and Consequences*. London: Sage Publications.

Gilmore, R. 2007. *Golden Gulag: Prisons, Surplus, Crisis, and Opposition in Globalizing California*. Berkeley: University of California Press.

Goffman, E. 1961 [1990]. *Asylums: Essays on the Social Situation of Mental Patients and Other Inmates*. New York: Doubleday Anchor.

Heins, V. 2005. Giorgio Agamben and the current state of affairs in humanitarian law and human rights policy. *German Law Journal* 6(1): 845-860.

Levy, C. 2010. Refugees, Europe, camps/state of exception: 'Into the zone', the European Union and extraterritorial processing of migrants, refugees and asylum-seekers (theories and practice). *Refugee Survey Quarterly* 29(1): 92-119.

Melossi, D. 2004. The cultural embeddedness of social control: Reflections on a comparison of Italian and North American cultures concerning punishment, in

Criminal Justice and Political Cultures: National and International dimensions of crime control, edited by T. Newburn and R. Sparks. London: Willan, 45-64.
Mesnard, P. 2004. The political philosophy of Giorgio Agamben: A critical evaluation. *Totalitarian Movements and Political Religion* 5(1): 139-157.
Mountz, A. 2004. Embodying the nation-state: Canada's response to human smuggling. *Political Geography* 23: 323-345.
Pager, D. 2007. *Marked. Race, Crime and Finding Work in an Era of Mass Incarceration*. Chicago: The University of Chicago Press.
Pallot, J. 2007. 'Gde muzh, tam zhena' (Where husband is, so is the wife): Space and gender in post-Soviet patterns of penality. *Environment and Planning A* 39: 570-589.
Peck, J. 2003. Geography and public policy: Mapping the penal state. *Progress in Human Geography* 27(2): 222-232.
Peck, J. and N. Theodore 2008. Carceral Chicago: Making the ex-offender employability crisis. *International Journal of Urban and Regional Research* 32(2): 251-281.
Programa de mejoramiento del sistema penitenciario colombiano, apéndice 11 al anexo al Acuerdo general para asistencia económica, técnica y otras asistencias relacionadas entre el gobierno de los Estados-Unidos y el Gobierno de la República de Colombia 2000/2001, Bogotá.
Rhodes, L. 2001. Toward an anthropology of prisons. *Annual Review of Anthropology* 30: 65-83.
Sentence T-153 1998. Constitucional Corte of Colombia.
Sibley, D. and B. Van Hoven 2008. The contamination of personal space: Boundary construction in a prison environment. *Area* 41(2): 198-206.
Wacquant, L. 1999. *Les prisons de la misère*. Paris: Raison d'agir.
Wacquant, L. 2002. The curious eclipse of prison ethnography in the age of mass incarceration. *Ethnography* 3(4): 371-397.
Wacquant, L. 2009. *Punishing the Poor: The Neoliberal Government of Social Insecurity*. Durham, NC and London: Duke University Press.
Walters, W. 2008. Acts of demonstration: Mapping the terroritory of (non) citizenship, in *Acts of Citizenship*, edited by E. Isin and G. Nielsen. London: Zed Books, 182-207.
Wilkey W. and G. Rivera 2002. Plan Colombia: A successful long-term effort. *Corrections Today*, 1 December.

Chapter 14
Poetic Testimonies of Incarceration: Towards a Vision of Prison as Manifold Space

Mason McWatters

Introduction

In recent years carceral geographers have turned toward examining the physical interiors of prisons to better understand how space becomes produced, contested, personalized and segmented within these institutional environments (Baer and Ravneberg 2008, Dirsuweit 1999, van Hoven and Sibley 2008, Martin and Mitchelson 2009, Moran 2011, Sibley and van Hoven 2009). Among the findings to emerge from these investigations is the thesis that prisons are not, or at least are no longer, the kinds of 'total institutions' that Goffman (1961) wrote about half a century ago. This is a 'non-totalizing' thesis about prison spaces that also has garnered a substantial amount of support from a number of criminologists and sociologists whose research explores carceral spaces, identities and experiences (da Cunha 2008, Farrington 1992, Fiddler 2006, Guilbaud 2010, Jewkes 2005).

Whereas totalizing views tended to regard prisons as institutional spaces that were absolutely spatially contained and socially insular, more recent, non-totalizing views have begun to reimagine prisons as dynamic spaces replete with temporal flows and social encounters that defy absolutist conceptions of prison as monolithic capsules of space and time. This paradigmatic shift is well illustrated by Baer and Ravneberg's (2008: 214) observational analysis of two European prisons, in which they argue that these institutions should be defined not 'as systems of social isolation or insulation with homogeneous and stable environments but as societal microcosms'. To help frame their argument the authors employ Foucault's (1998) concept of a heterotopia in order to illustrate how modern prisons are incompatible juxtapositions of heterogeneous spaces in which boundaries between inside and outside frequently become blurred or inverted. I contend that non-totalizing revisions of prison space, such as this, signify a positive step toward better understanding how individuals actually inhabit institutional spaces.

In this chapter, I wish to add to understandings about how prison space is actually experienced by those for whom it is an ordinary space of daily life: its inmates. To do this I shall draw upon poetic testimonies, written during the latter half of the twentieth century by prisoners incarcerated in the United States, in order to explore a variety of ways in which prison as an inhabited space is represented and shaped through creative writing. Based on this poetic analysis, I

find that the lived space of prison is much more plastic, fluid and manifold than totalizing notions permit. As such I argue in support of efforts to expand the spatial imaginary of lived spaces of incarceration. However, importantly, I argue that as we expand toward new possibilities beyond the total institutional framework, we must not lose touch with the fact that totalizing experiences of prison and incarceration still remain important possibilities as well. Indeed, in some instances prison remains just as totalizing as ever.

Chapter Overview

In the following section I shall turn to explore several different possibilities for prison space which include, but also exceed, totalizing experiences. Positing that there is a broad spectrum of possible lived prison spaces, I wish to present three common representational 'sites' along this spectrum where poetic testimonies frequently intersect and overlap. I regard these three common sites of experience as generic ('genre-like') locations which crudely plot out two extreme and one intermediate possibility for lived prison space (Table 14.1). While these spatial genres are in no way intended to represent the entire spectrum of incarcerated experiences, whose dynamicity surely exceeds the limits of representation altogether (Thrift 2008), I argue that they accomplish two important aims. First, these spatial genres affirm a variety of experiential textures and affective patterns which are discernible across broad corpuses of poetic testimony about incarceration. Affirmation of the manifold textures and patterns of prison space and prison experience needs to be done if we are to advance beyond totalizing notions of incarceration. Second, and responding to this demand, I contend that these three spatial genres offer several evenly-spaced 'starting places' for exploring further lived possibilities of being incarcerated.

Table 14.1 Spatial genres of lived prison space

Spatial Genre	**Topic**	**Heterotopic**	**Atopic**
Temporality	Fixed, synchronic	Fragmented, diachronic	Atemporal, infinite
Meaning/Identity	Definite, essential	Contested, paradoxical	Indefinite, empty
Representationality	Transparent, representable	Translucent, intertextual	Opaque, non-representable
Quality	Real, proper	Surreal, incongruent	Unreal, improper
Logic	Logical, literal	Analogic, abstract	Alogical, nihilistic
Legibility	Relatively legible	Semi-legible	Relatively illegible

The first genre is prison as topic space: a transparent, representable and totalizing place of enclosure about which inmates express an absolute sense of existential confinement, oppression and dispossession. According to this representational theme, the meaning of prison as an absolute space of deprivation is coherent and clearly legible. The second genre is prison as heterotopic space: a multitude of incompatible spaces that contradict the former notion of prison as a coherent and absolute place. In representations of prison as heterotopic space, the absolute divisions between inside/outside, power/powerlessness and possession/dispossession dissolve into a juxtaposition of unresolved, heterogeneous senses of space. Moving along the spectrum of prison experience from topic to heterotopic genres of space, the sense of prison as an absolute place fragments into a multiplicity of different spaces of sense, or senses of space, that undermine the idea of prison as a total institution. The third genre is prison as atopic space: literally a non-representable space whose sense is absolutely confounding and improper. This is a disorienting experience of prison as a groundless and placeless space which lacks meaning or sense.

In the discussion that follows this section, I argue in support of a spatial theory of prison which acknowledges the 'contemporaneous plurality' (Massey 2005) of space within carceral institutions. Thus, I envision these three spatial genres as constituting a spectrum of prison experience in which senses of incarceration are dynamic, manifold and always in flux both within and between incarcerated subjectivities. Here I also briefly consider how practices of creative writing might be employed by subjectivities as a mediating strategy for enacting new and better possibilities for their lived environments of incarceration.

In the conclusion, I argue in support of a movement within carceral geographies to transcend conceptualizations of prison as totalizing spaces of absolute isolation, containment and dispossession. However I also caution that we must be careful not to allow new possibilities for prison space to be fragmented, dynamic and heterogeneous to crowd out still-extant possibilities for prison space to be absolutely oppressive or senseless in certain instances. Indeed, I argue that we must not become too enamoured with any one way of conceptualizing institutional spaces, as the grounded experiences of these spaces are always multiple and beyond singular representation. As such, I suggest that scholars of incarceration in geography and elsewhere strive for complex conceptual framework about prison space as plurally topic, heterotopic and atopic.

Three Genres of Prison Space in Poetic Testimonies

To inform my analysis I draw upon two anthologies of prison poetry composed by individuals incarcerated in modern American prisons. Amidst a rich and diverse literary tradition of prison writing, I have selected these two anthologies in particular because each offers poetic testimonies from an impressive variety of authors: 78 in total. This authorial variety is appealing for my analysis because

it offers a wide range of different sites, contexts and positionalities for thinking prison space in terms of its variable possibilities.

The first anthology is *Captive Voices: An Anthology of Literary Works by Folsom Writers*, published in 1975, which contains the poetry and short stories of 18 male authors of various ethnicities incarcerated in the eponymous Californian state penitentiary (*Captive Voices* 1975). The second anthology is *The Light from Another Country: Poetry from America's Prisons*, published in 1984, which contains the poetry of 60 authors – both male and female of various ethnicities – imprisoned in various state penitentiaries across the United States (Bruchac 1984). From these anthologies, my analysis includes those poems in which the experience of incarceration and/or the existential space of prison itself occupies a central place in the author's poetic testimony.

Genre I: Prison as Topic Space

The first experiential genre I wish to discuss is prison as topic space. Topic space, as I define it, exhibits qualities of place, being a relatively stable environment that conforms to rational notions of Cartesian space. Topic space is logical and literal; an identifiable, holistic place constituted by a grounded and synchronic here and now.

In the context of incarceration, topic space conforms to common representations of prison as a total institution, which Goffman (1961: 1) characterized as a place 'cut off from the wider society' in which 'the various enforced activities are brought together into a single rational plan purportedly designed to fulfil the official aims of the institution' (ibid: 6). This topic notion of prison as a total institution conforms to popular representations of prison as a site of absolute confinement, deprivation and oppression.

This is the very image of the prison space encountered in 'Feeling Locked In', written by a California inmate named Willi X:

FOLSOM
FOLSOM
FOLSOM
You hold me
you hold my hate
with your walls
you keep out love
you keep out life
with your bars.[1]

In this poem the raw feelings of prison as a place of utter dispossession leap from the page. It connotes a cold, concrete space that is restrictive to the point

1 *Captive Voices* 1975: 211.

of suffocation. This is similar to the sense conveyed in the poem 'Real Deal Revelation', by Raymond Ringo Fernandez, who was incarcerated in Attica Correctional Facility in New York at the time he wrote the following:

Aqui
you can't loosen up
you can't say
certain things
or look at people
like you care
…
you can't remember
your name sometimes
you forget what your father looks like
you don't remember
the last time
you heard momma cry

Aqui
you can't get loose
can't get loose
can't
be yourself
can't …[2]

This is a topic account of prison as a place that strips one of his humanity to the point where one's entire world becomes limited to the constricting walls of the prison itself. It evokes a sense of total deprivation in which the confines of prison are defined negatively as an absolute space of inhumane exception. In this topic space of destitution one is deprived of all transcendent facets of identity beyond the walls. As a topic space, prison only accommodates the bare identity of embodied prisoner in the destitute space-time of the here and now.

This here and now is a highly regulated space-time of banal, cyclical repetitions, as Daniel Berrigan's poem 'Rehabilitative Report: We Can Still Laugh' connotes:

In prison you put on your clothes
and take them off again.
You jam our food down
and shit it out again
You round the compound right
to left and right again.
The year grows irretrievably old

2 Bruchac 1984: 98.

so does your hair burn white.
The mood; one volt above
one volt below survival,
roughly per specimen, space
sufficient for decent burial.[3]

As this poem demonstrates, the rigid structure of topic imprisonment extends to time as well as space. Nothing seems to change in the topic space of prison. The walls confine, time repeats itself, or else drags on slowly, and a sense of prison as an absolute place of deprivation becomes unmistakably evident.

Genre II: Prison as Heterotopic Space

The second spatial genre represented in the corpus of poetic testimonies surveyed for this analysis is prison as heterotopic space. Moving along the spectrum of prison experience from topic to heterotopic genres, the absoluteness of prison as a totalizing space confined to the here and now begins to crumble and dissolve into an array of heterogeneous spaces and times. In heterotopic space, experience becomes fragmented to the point that absolute clarification about prison as a singular, fixed place becomes impossible. In heterotopic space, the fixed boundaries between real and imaginary, literal and figurative, and interior and exterior become blurred as space acquires a sense of ecstatic fluidity that is unbounded, decentred and sometimes even liberating.

Foucault (1998, 2002) originally wrote of heterotopias as spaces in which incompatible juxtapositions are brought into (dis)alignment in such a way as to destroy their syntactical order and coherency. Foucault's heterotopias are 'monstrous' spaces in that they bring paradoxical arrangements into being in ways that seem *un*real. They are 'disturbing…because they make it impossible to name this *and* that, because they shatter or tangle common names, because they destroy "syntax" in advance, and not only syntax with which we construct sentences but also that less apparent syntax which causes words and things (next to and also opposite one another) to "hold together"' (Foucault 2002: xix, cited in Barnes 2004: 574).

In geography, Hetherington (1997) has written masterfully about the unsettling effects of heterotopias, which he describes as spatial assemblages of distributed materialities, events and social practices. On this he writes:

> Heterotopic relationships unsettle because they have the effect of making things appear out of place. The juxtaposition of the unusual creates a challenge to all settled representations; it challenges order and its sense of fixity and certainty. This process of similitude is revelatory, like a collage, it brings forward the out-of-place and offers it up as a basis for alternative perspectives and orderings, revealing what is hidden among the ruins: little fragments of past, forgotten

3 Bruchac 1984: 47.

> lives, found objects, strange, unsettling novel things that have a poetic wonder about them. (Hetherington 1997: 50)

Heterotopic sense has this dynamic and, dare I say, cinematic quality to it. It is devious in the way that it veers off course from the ordinary, breaking – or at least suspending – the rules concerning what is possible, logical and proper. In this regard, elements of difference are critical to heterotopic senses of the world. Heterotopic space is textured with a temporal movement of digressions, rearrangements and unsettling (un)becomings. This diachronic motion is what gives the concept of heterotopias such purchase and possibility within geography.

These dynamic qualities make heterotopias an appropriate analytical tool for exploring carceral geographies beyond absolute notions of totality. This is the very point that Baer and Ravneberg (2008) make in their analysis of prison spaces in Norway and England, in which they argue that distinctions between inside and outside appear less-than-total. Reflecting on their observations, they write:

> In the process, we noticed multiple, simultaneous distinctions and indistinctions that appeared to us to be part of the prison environment. Distinction and indistinction may seem contradictory, but there are many contradictions in everyday life within and outside the prison; the concept of heterotopia allows for this through its principle of incompatible juxtaposition. (Baer and Ravneberg 2008: 212)

My analysis supports this claim about the unexpected and incongruous qualities of lived carceral spaces.

One of the ways that the topic stability of prison as a space of total destitution is unsettled is through prisoners' re-inscription of their situated identities in ways that subvert expectations about their utter lack of agency in and control over their environments. This rewriting of one's identity *vis-à-vis* prison space is what Leon Baker – a man who, not incidentally, voluntarily spent the majority of his prison term in solitary confinement – performs in his poem 'Getting Back to Work (in solitary confinement)':

> For four months the guards sabotaged
> my efforts, confused my thoughts, destroyed
> my typewriter and had me surrounded by:
> degenerates, whose meaningless chat, chat, chat
> choked my poetry damn near to death.
> …
> Getting back to work,
> with a brand new poetic rap, pen in hand,
> note book across my lap, and rats running
> over my floor (in solitary confinement).[4]

4 Bruchac 1984: 35.

In Baker's testimony the trope of prison as a space of destitution remains. However in heterotopic fashion this destitution is inverted so that it is not social isolation and spatial confinement that create feelings of dispossession, but rather socio-spatial crowding and overbearing distractions which keep the writer from his work. In this heterotopic inversion, solitary confinement paradoxically becomes a place of refuge where Baker can perform his real identity as, first and foremost, a writer, not a prisoner. In this sense, heterotopic space serves as a site of otherness in which subjectivities are able to create space for performances of alterity.

In certain instances heterotopic space provides a destabilized context in which inmates are able to challenge the supposed powerless nature of their positionalities by disrupting the fixed structure of the prison's social order. By contesting their identities as powerless prisoners, these individuals not only perform new subjectivities like Baker's purpose-driven writer persona; in the process they also 'unsettle the flow of discourse' (Hetherington 1997: 43) that would represent prison as an absolute place of dispossession. This is precisely what Philip Brasfield, who is serving a life sentence in Texas, accomplishes in 'An Interview', in which he recounts a visit with the warden to discuss his subversive writing:

> The warden reads my mail and frowns.
> …
> His office smells of turpitude.
>
> I'm told to read aloud those words
> And hear the heaving whiskey-breath:
> Outraged, as a smile escapes my lips.
>
> Eyes locked in struggle, he looks away
> Dismissing me to my imprisonment
> Of thinking free.[5]

Here the topic coherency of the prison/prisoner as an absolute space/subject of dispossession truly becomes disrupted as the sure distinctions between empowerment/disempowerment, knowledge/ignorance and insideness/outsideness are disassembled and subverted. The topical syntax of the total institution breaks down into heterotopias and heterotexts that simultaneously tell multiple place narratives about prison space. A fugitive smile simultaneously hides and reveals another space of power and knowledge in which the roles and relations are more complex than they appear on the surface.

In heterotopic spaces of prison, proper distinctions between observer/observed and subject/object break down in ways that defy the panoptic model of institutional power. In 'On Youth, The Warden & Solitary!' Baker performs the feeling of

5 Bruchac 1984: 63.

devious vertigo as sites of knowledge are deterritorialized and reterritorialized back and forth, back and forth, across the axes of blinds and bars:

> While the warden peeps through the venetian
> blinds of my eyes and tries to learn the
> secrets held firm by gray-matter in my
> skull, and wonders why there are no age cob-webs
> around the sockets which house optic nerves
> leading to an outside world of his man-made darkness, I laugh!
> …
> Peeping at him, peeping at me, peeping at him
> straining his eyes, trying to see beyond the
> darkness he created. Peeping at me, peeping
> at him, peeping at me enjoying the rewards
> of being without age-spots.[6]

The further one ventures into the heterotopic spaces of prison, the more difficult it becomes to distinguish interiority from exteriority. Indeed, as Baer and Ravneberg (2008: 208) remarked in their analysis, 'heterotopias open and close in such a way that they are both isolated and penetrable at the same time'. Furthermore, as Hetherington (1997: 34) writes, 'heterotopias are spaces in which an alternative social ordering is performed' in such a way that strict limits and controls become fluid and porous.

However, recalling that heterotopic space is a relational space, we must acknowledge that it is not only the spatialities of here and there that become distorted, but also the temporalities of now and then. Thus, prison as heterotopic space poses challenges to the embodied, synchronic limits of the here and now, imposed by the sheer physicality of the prison walls. In many instances this is achieved by way of transcendent performances which ferry prisoners beyond the topic space of deprivation into spaces and times beyond incarceration. Consider, for example, what something as seemingly inconsequential as music has the power to achieve in Yasmeen Jamal's poem, 'All That Jazz':

> I drift to another time and space when
>   jazz listens to my moods.
> Sometimes Mose Allison and Miles Davis
>   come with me.
> Morgana King and Yusef Lateef know where
>   I've been and where I'm going.
> They've been there too.
> I drift to another time and space when
>   jazz listens to my moods.

6 Bruchac 1984: 34.

...
I am out there,
Laid all the way back.
Driftin' to another time and space when
  jazz listens to my moods.
And I mean I leave this space.
I trip with no help.
Ask Art Tatum,
He watches me.[7]

Jamal's insistence of the power of music as a 'potent mediator' (Latour 2005) denies us, as analysts, the ability to casually dismiss the claims of her testimony to be merely a metaphor for the desire to escape. Indeed, I argue that this *is* escape from the topic mania of institutional deprivation. In granting poetic expression the power to generate space-times of alterity, I agree with Bachelard's (1994: 150) pronouncement that 'to get out of prison all means are good ones'. Even if escape only entails checking out of the topic space of prison for a while to drift into an alternate space-time where jazz feels one's pain, this escape clearly matters in an affective, material sense.

As spaces of alterity, heterotopias 'bring together heterogeneous collections of unusual things without allowing them a unity or order established through resemblance. Instead, their ordering is derived from a process of similitude which produces, in an almost magical, uncertain space, monstrous combinations that unsettle the flow of discourse' (Hetherington 1997: 43). One of these discourses that the genre of heterotopic space shatters is the notion that prisons are so socially alienated and spatial isolated that they literally exist outside of 'ordinary' social space. Rather, what the heterotopic spaces of prison demonstrate through juxtaposition is the dissolution of clear boundaries between inside and outside, contact and isolation. This dissolution is masterfully articulated in Jimmy Santiago Baca's poem, 'I am Sure of It'. In this poem Baca – who, at the time, was incarcerated in New Mexico on a drug charge – wanders through an extensive, stream-of-conscious journey out into an extensive world of heterotopic space:

I take a few steps to the toilet and pull myself up,
my mouth on the back grill at the back of my cell,
just about to holler down to my buddy in another cell
when I am struck silent by the window across from me,
and I look outside, upon a few convicts at dusk,
running in pairs around the baseball diamond,
...
The grass is green, trees lulled in deep spring slumber,

7 Bruchac 1984: 147.

sun going down at the west edge of earth.
…
The smell of cool spring is in the night air, above all things,
and toy-sized cars crawl in black distances
...
Grass crickets tune their fiddles on the wind;
by rumbling trucks on the freeways,
construction workers coming and going into homey bars,
greasy caps cocked to one side, smoking cigarettes,
stroking pool cues through gritty thick fingers,
suck their unbrushed teeth watching intently,
as Tammy Wynette wails out her country hits
to dimes and quarters plunked in the jukebox,
among the bearish beer-drinking, smoke swirling workers,
coughing and cursing and smiling,
while their fathers push on screen doors,
enter yellowed kitchens and drink coffee with married daughters
…
All the while we convicts live in a smooth block of rock,
convicts scan through their TV guides,
and through all the different blubbering channels,
an old Mexican song floats out from someone's cell,
recalling memories; under the screams and gunshots
tumbling out of TVs, memories float heavy in me,
and I think of tennis shoes my grandfather
bought for me as a child for my birthday.

My life so filled with simple things!
With beds and people crying and laughing and fighting,
towns and voices and kisses and unforgettable nights,
walking on sidewalks or through grass at dusk,
this is life, I am sure of it…
thinking all the time, this is life
even in prison, respecting each other, helping each other,
close or far away, it doesn't matter, I am sure of it.[8]

In Baca's poem the prison is anything but a total institution. Rather here and now, there and then, swirl into a 'compressed mélange of seemingly incompatible juxtapositions' (Baer and Ravneberg 2008: 213). Surfaces flow from concrete to grass, then back to concrete; from bar top to table top, to block of rock. Scales of human-like proportion reduce down to scales of insect-like proportion, which then inflate to scales of God-like proportion. The now of doing time morphs into

8 Bruchac 1984: 28-29.

the then of other times, good times and once-upon-a-times. One becomes many. Interiority opens onto exteriority as scenes, subjectivities, texts, times and spaces fall into turbulent assembly momentarily gaining quasi-coherency as a heterotopic microcosm of the busyness of being human.

The farther we venture into these heterotopic spaces of prison, the less absolutely prison appears as a totalized place. Thus, the image of prison as a topic space not only becomes fragmented *intra*textually, like in Baca's poem above, but also *inter*textually, as singular testimonies begin to accumulate into a collective corpus. These heterotopic processes shatter generalities about prison as absolutely totalizing in every instance. Through these processes the integrity of prison as a topic space fragments into a semi-legible, heterotopic plurality of space-times. If we continue moving in this direction towards disintegration on the spectrum of prison experience, we reach a threshold in which the multiple, relative, contested meanings of prison eventually fragments into an atopic lacuna of incoherent placelessness.

Genre III: Prison as Atopic Space

In atopic space, the logics of being-in-place melt away into infinite disorientation. In atopic space, prison becomes nowhere, which is indistinguishable from everywhere else. Atopic spaces are improper spaces of sense beyond sense (Nancy 1997). In the context of prison, atopic spaces are like black holes in which subjectivities drift into nihilistic despair, as Folsom inmate Carson H. Cook performs in 'Is This Paradise':

> I can't see pollution, waste, destruction or greed!
> I can't see crime, death or pain!
> I can't see hunger or want!
> I can't see![9]

As one journeys through this poem, sense and meaning come undone. Presence becomes reduced to an atopic void of space in which nothing is apparent or sensible. Prison literally becomes a senseless space. The paradox of prison as atopic space is that it is simultaneously complex and bare. In this space, proper feelings of self-possessed emotions give way to improper feelings of diffuse affects that can only be framed negatively from without, as an absence of sense, or sense of absence. This is precisely the non-representational effect of atopia that Michael Knoll struggles to relate in 'Prison Letter':

> You ask what it's like here
> but there are no words for it.
> I answer difficult, painful, that men

9 *Captive Voices* 1975, 47.

die hearing their own voices. That answer
isn't right though and I tell you now
that prison is a room
where a man waits with his nerves
drawn tight as barbed wire, an afternoon
that continues for months, that rises
around his legs like water
until the man is insane
and thinks the afternoon is a lake:
blue water, whitecaps, an island
where he lies under pale sunlight, one
red gardenia growing from his hand –

But that's not right either. There are no
flowers in these cells, no water
and I hold nothing in my hands
but fear, what lives
in the absence of light, emptying
from my body to fill the large darkness
rising like water up my legs:

It rises and there are no words for it
though I look for them, and turn
on light and watch it
fall like an open yellow shirt
over black water, the light holding
against the dark for just
an instant: against what trembles
in my throat, a particular fear,
a word I have no words for.[10]

Prison becomes nothing: a word there are no words for. Upon each attempt to say what this lived space is like, Knoll falters to capture its excessive sense beyond sense. The experience of prison space becomes uncertain and representation falters and falls short (Harrison 2007).

Derrida (1993) and Agamben (1999), among other continental theorists, would refer to this kind of troubling and improper affect that exceeds the language of facts and verification as an aporia. A fitting definition of the term, albeit from the mid-seventeenth century is provided in the *OED*: 'Aporia is a figure whereby the Speaker sheweth that he doubteth, either where to begin for the multitude of matters, or what to do or say in some strange or ambiguous thing.' This is exactly the doubt Knoll conveys in struggling to describe prison as atopic space.

10 Bruchac 1984, 184.

In extreme instances such as this, prison is no more and no less than a groundless, infinite space of disorientation.

Discussion

Variables of Experience in Incarceration

Even in prison, space is never static. The excess of prison as atopic is eventually bound to crystallize into something definite. This is, indeed, what we witness in Folsom inmate Javier Viramontes' poem, 'baby LOVE':

> I am lost now, lost within the empty shadows of eternity
> Where drowned passions float and death has left a scree,[11]
> And where counterfeit people flout
> From the wings of a single shout…
>
> Where am I? someone has plucked the moon from out its flight,
> The stars have left no trace upon the sky,
> These nights of wrest have broached their light,
> And left a river deep and cold.
> Styx? perhaps, I cannot say,
> For all I see is stone: on cold cement my head I lay.[12]

From the abstract void of atopic space, of everything and nothing, Viramontes is ferried back to the present concreteness of cold stone. The groundlessness of prison as atopic space gives way to a topic space of concrete deprivation, as Viramontes returns to the grim reality of 'all I see': everything, only this.

Based on the foregoing analysis, the argument I want to make is that prison space is not one place, many places or no place at all, but rather all of these variable topic, heterotopic and atopic spaces across spans and spectrums of experience that are both intrasubjective and intersubjective in nature. Once we take into consideration how space is plastic and manifold even in carceral settings, we see that the experiences and meanings of being incarcerated are always multiple and in flux. What feels like a topic space of deprivation in one instance, situation or context is bound to vary, differ and defer across other instances, situations and contexts. The genres of topic, heterotopic and atopic spaces that I have presented above are meant to suggest three common possibilities among the spectrums and spans of prison experience in the contemporary American context.

---

11 Scree is a mass of detritus, forming a precipitous, stony slope upon a mountain-side.

12 *Captive Voices* 1975, 158.

The notion that lived spaces of incarceration vary across different geographical contexts is relatively intuitive. What is less intuitive but just as significant is the idea that lived spaces of incarceration also have the dynamic capacity to vary across different historical situations and eventful contexts in situ. At one moment perhaps the affective despair of isolation, or the chance witnessing of random violence, come to define prison as no place at all; a senseless lacuna of inhumanity. Nothing makes sense in this instance of space-time. In another instance perhaps rumours spread through air ducts, giving inmates privileged knowledge that guards do not have; or perhaps inmate and guard pause to talk about life, family or current events as equals. In these moments prison becomes an incompatible juxtaposition of subjectivities and spaces. Yet there will also always be moments of experience when prison is simply nothing more than a place of absolute deprivation: four cold walls. Perhaps these are feelings of confinement that an hour of fresh air, a glimpse of the world beyond or a brief visit from family only serve to heighten by way of contrast. Or perhaps not. Yet, all things must pass; even this feeling of prison as nothing more than a place of deprivation. Thinking incarceration and carceral environments in terms of ranges and fluctuations between topic, heterotopic and atopic variables of space is one strategy for better accounting for this dynamicity.

My suggestion is that prison, like any other meaningful dwelling place, is not a static or uniform space but rather a transitive and manifold range of space-times with topic, heterotopic and atopic variables of experience. Some days prison is all there is: a suffocating totality; other days prison is tolerable: a piece of a greater, incompatible world. In certain instances it is peacefully quiet; in other instances it is brutally violent. Meanings, feelings and experiences vary – even in prison. Some days a letter arrives bringing with it traces of a world beyond. Other days a banal sense of sameness makes transcending the closed world behind bars impossible. In any case what is critical to note here is that these manifold spatial possibilities represent not just a variety of introspective, subjective perspectives of the same 'objective' space. Rather they represent real mediations of sense in and with one's environment through which different material and affective varieties of space become produced.

Poetic Interventions

Within the context of these manifold possibilities for prison space to become in every instance, what I suggest also merits further attention are considerations for how creative writing might serve as a powerful mediator that prisoners may use to actively realize certain spatial possibilities for their lived environments. In other words, given the variables of experience explored above, I would suggest that we consider what mediating strategies and technologies inmates have at their disposal to affect certain spatial outcomes for their incarceration.

Undoubtedly, the creative practice of writing, especially poetic writing, is critical to the repertoire of tools inmates have ready-to-hand for shaping and reshaping their lived environments. In affective environments which are often

highly steeped in violence, inhumanity and deprivation, writing becomes a creative intervention for shaping new possibilities for one's incarceration. For instance, in reference to the adage that 'poetry can help make us human', Pennsylvania native Jack L. Anderson writes the following about his incarceration:

> Personally, I have found that simple statement to be of tremendous, perhaps even profound importance, what with my being confined in a State Correctional Institution (i.e.; Prison) at the time..., and it being a place of such a nature that the official custodians found the MASTER/slave theme a very great deal to their liking. Thus; from that time on, throughout everything, Poetry became both vehicle and destination, transcendence and transformation, the hope and the promise, ultimately, the medium capable of sustaining both out-cry and lingua franca. (Bruchac 1984: 14)

In this way, the creative potential of poetic language becomes 'a way of sledge-hammering these walls' (Chevigny 2005: 247) to reshape one's lived context in less oppressive terms.

In addition to being a powerful political mediator, writing in contexts of incarceration may also become a valuable therapeutic practice for reshaping one's existential place in the world. This transformative, therapeutic potential is evoked in Jimmy Santiago Baca's recollection of his initial attempts at creative writing upon being incarcerated:

> Until then, I had felt as if I had been born into a raging ocean where I swam relentlessly, flailing my arms in hope of rescue, of reaching a shoreline I never sighted. Never solid ground beneath me, never a resting place. I had lived with only the desperate hope to stay afloat; that and nothing more.
>
> But when I wrote my first words on the page, I felt an instant island rising beneath my feet like the back of a whale. As more and more words emerged, I could finally rest: I had a place to stand for the first time in my life. (Chevigny 2005: 254)

In this sense, writing may become a powerful tool of intervention for reshaping prison as an atopic space of disorientation into a more habitable and grounded place of dwelling.

Writing, and communication in general, possess the creative potential to make and reshape spaces of existence. As Adams (2009: 167) has so elegantly put it, 'communications do not simply reside within pre-established and finished places; instead they contribute to creating places'. In contexts of incarceration, writing does not merely represent space in a passive and secondary sense; rather writing represents an active intervention for composing new materialities of prison space and for performing new 'more-than-incarcerated' subjectivities.

Imagining Prison Space

What we are left with is two different but compatible perspectives of prison space. The first is prisons as always heterotopic spaces of incompatible juxtapositions on a trans-subjective, intertextual level. This heterotopic representation of prison space is one that acknowledges the infinite diversity of singular experiences of 'doing time' across different contexts. It also maintains a positive tension between signifier and signified so that no final meaning about prison can ever be reductively generalized. Quite simply, at the trans-subjective level, prison space is always a heterotopic abstraction of contradictory juxtapositions in which lines of power and containment are never fully clear or finalized. In prison as heterotopic space, rather, these lines are always entangled, transforming, blurred and contested (Sharp et al. 2000). In this space, prison is not apart from society but rather embedded within it. Thus, based on the foregoing analysis, I want to affirm the argument that prisons be studied as heterotopic spaces.

However, as much as prison holds as heterotopic space on a general level it can become problematic as soon as we shift scales to consider singular experiences of incarceration. The problem becomes this: if we are to consider prison as inherently heterotopic and 'not-so-total' (Farrington 1992), what does this do to possibilities for prison to still be absolutely (a)topically totalizing in certain instances? Quite simply it risks denying possibilities that sometimes, in certain instances, prison is the absolute all of one's existence. Remaining open to pluralities, ranges, spans and spectrums of possibility for prison space in a way that no one possibility overrules the others is what testimonies of being incarcerated urge for carceral studies.

Conclusion

Based on the poetic testimonies presented here, I contend that we must take care to attend equally to the manifold qualities and possibilities for prison space. Testimonial accounts of incarceration resonate to a great degree with notions of prison as heterotopic spaces in which lines of interiority and exteriority, and power and powerlessness are blurred and entangled. However, for as much appeal as this heterotopic genre offers for better understanding institutional spaces across a variety of scales, there are still other genres and other possibilities for prison space, including totalizing ones, which we cannot afford to neglect. Attending to prison space in terms of dynamic spectrums and spans of manifold possibilities that all deserve our attention is what needs to be undertaken.

Furthermore, with this variability of experience in mind, I suggest that carceral studies embrace a move to more explicitly consider how incarcerated subjectivities employ certain mediating strategies and technologies to actively reshape their existential contexts of incarceration. Creative writing, as I have suggested, constitutes one important practice for redefining existential spaces of incarceration in ways that are more politically and therapeutically tolerable. What other kinds of

mediating strategies and technologies might also belong to prisoners' repertoires for realizing new spatial and environmental possibilities for their incarceration?

Acknowledgements

I would like to extend my gratitude to all the talented poets whose work appears in this chapter. I would also like to acknowledge Joe Bruchac for his permission to reprint the poetry appearing here. Thank you also to Paul Adams, Bell Chevigny, the editors of this volume, and the two anonymous reviewers for their insightful comments. Any errors are my own.

References

Adams, P.C. 2009. *Geographies of Media and Communication: A Critical Introduction*. Malden, MA: Wiley-Blackwell.

Agamben, G. 1999. *Remnants of Auschwitz: The Witness and the Archive*. New York: Zone Books.

Bachelard, G. 1994. *The Poetics of Space*. Boston: Beacon Press.

Baer, L.D. and B. Ravneberg 2008. The outside and inside in Norwegian and English prisons. *Geografiska Annaler, Series B*, 90(2): 205-216.

Barnes, T. 2004. Placing ideas: Genius loci, heterotopia and geography's quantitative revolution. *Progress in Human Geography* 28(5): 565-595.

Bruchac, J. (ed.) 1984. *The Light from Another Country: Poetry from American Prisons*. Greenfield Center, NY: The Greenfield Review Press

Captive Voices: An Anthology of Literary Works by Folsom Writers. 1975. Paradise, CA: Dustbooks.

Chevigny, B.G. 2005. 'All I have, a lament and a boast': Why prisoners write, in *Prose and Cons: Essays on Prison Literature in the United States*, edited by D.Q. Miller. Jefferson, NC: McFarland, 246-271.

da Cunha, M.I. 2008. Closed circuits: Kinship, neighborhood and incarceration in urban Portugal. *Ethnography* 9(3): 325-350.

Derrida, J. 1993. *Aporias*. Stanford, CA: Stanford University Press.

Dirsuweit, T. 1999. Carceral spaces in South Africa: A case study of institutional power, sexuality and transgression in women's prison. *Geoforum* 30: 71-83.

Farrington, K. 1992. The modern prison as total institution? Public perception versus objective reality. *Crime and Delinquency* 38(1): 6-26.

Fiddler, M. 2006. The penal palimpsest: an exploration of prison spatiality. Unpublished doctoral dissertation. Keele: Keele University.

Foucault, M. 1998. Different spaces, in *Aesthetics, Method and Epistemology*, edited by J.D. Faubion. New York: New Press, 175-185.

Foucault, M. 2002. *The Order of Things: An Archaeology of the Human Sciences*. London: Routledge.

Goffman, E. 1961. *Asylums: Essays on the Social Situation of Mental Patients and Other Inmates*. Garden City, NY: Anchor Books.

Guilbaud, F. 2010. Working in prison: Time as experienced by inmate-workers. *Revue française de sociologie* 51: 41-67.

Harrison, P. 2007. 'How shall I say it…?' Relating to the nonrelational. *Environment and Planning A* 39: 590-608.

Hetherington, K. 1997. The Badlands of Modernity: Heterotopia and Social Ordering. New York: Routledge.

van Hoven, B. and Sibley, D. 2008. 'Just duck': the role of vision in the production of prison spaces. *Environment and Planning D* 26(6): 1001-1017.

Jewkes, Y. 2005. Loss, liminality and the life sentence: Managing identity through a disrupted lifecourse, in *The Effects of Imprisonment*, edited by A. Leibling and S. Maruna. Cullompton: Willan Publishing, 366-388.

Latour, B. 2005. Reassembling the social: An introduction to actor-network-theory. New York: Oxford University Press.

Martin, L.L. and M.L. Mitchelson 2009. Geographies of detention and imprisonment: Interrogating spatial practices of confinement, discipline, law, and state power. *Geography Compass* 3(1): 459-477.

Massey, D. 2005. *For Space*. London: Sage.

Moran, D. 2011. Between outside and inside? Prison visiting rooms as liminal carceral spaces. *GeoJournal* [Online] Available at http://www.springerlink.com/content/p805836842373543/?MUD=MP [accessed 21 May 2012]

Nancy, J-L. 1997. *The Sense of the World*. Minneapolis: University of Minnesota Press.

Sharp, J.P., P. Routledge, C. Philo and R. Paddison (eds) 2000. *Entanglements of Power: Geographies of Dominance/Resistance*. London: Routledge.

Sibley, D. and B. van Hoven 2009. The contamination of personal space: Boundary construction in a prison environment. *Are*, 41(2): 198-206.

Thrift, N. 2008. *Non-Representational Theory: Space, Politics, Affect*. New York: Routledge.

Chapter 15
The Politics of Carceral Spectacle: Televising Prison Life

Jennifer Turner

Introduction

On Monday 5 September 2011, UK Justice Secretary Kenneth Clarke released an article in *The Guardian* newspaper holding the 'legacy of the broken [British] penal system' accountable for the actions of a 'feral underclass' during the London riots of August 2011. What was striking was that although Clarke described the criminals with highly charged terminology, he held *his* penal system to blame for the circumstances that led to this riotous behaviour. Clark claimed that over 75 per cent of those aged 18 or over who were brought before the courts had a prior conviction, forcing him and the government to think drastically about not just penal rehabilitation, but education, welfare and family policy, all of which can have serious impacts on propensity to commit crime.

Clark's rhetoric is but one example of how prisons have come to be seen not as separate, peripheral sites, but as windows onto, or even organizing principles of, modern social, political and, increasingly, economic orders. The intricate relations that connect prisons and their occupants to the 'outside' have been noted, for example, in the Million Dollar Blocks project based at the Justice Mapping Centre at Columbia University,[1] which identifies areas where states spend in excess of a million dollars a year to incarcerate the residents of single city blocks, firmly placing prisons in the context of housing policy and state budgetary priorities more generally. Reports on gang activity point to strong links between incarcerated gang members and those on the outside, and suggest that prisons are instrumental as recruiting stations (Spergel 1990), while religious groups, as well, find rich sources of converts within prison walls (Johnson 2004).

The principal aim of this chapter is to discuss this relationship between prison and society using notions of visibility and punishment in contemporary media, with particular reference to how the 'spectacle' of the incarcerated subject operates within television. With the abolition of the spectacle of visible punishment, such as public hangings, trial and torture, the penal system vanished behind prison walls (Foucault 1977/1991: 231), a displacement that served to nurture all manner

1 See (Graduate School of Architecture Planning and Preservation of Columbia University New York 2006).

of popular imagery concerning the nature of prison life (Turner, forthcoming). That is, although there was a physical distancing of the criminal from the rest of society via incarceration, as well as limited access to prisons, concerns over the criminal body were displaced into other arenas such as architecture and literature. As television increased in its popularity, so the prison programme replaced the public gallows of old (Mason 2000a). Criminals are now hidden behind the walls of the closed world of the prison, but the apparatus of penal mechanisms still remains at the forefront of popular geographical imaginations.

Although there has been research into both prisons (Gilmore 1999, 2007, Gregory 2006, 2007) and the constructions of 'otherness' in television representations (Horschelmann 1997), there is a gap in the literature regarding the specific construction of a carceral spectacle and subsequent identity constructions. One explanation for this, Mason observes, 'may be that programmes concerning prison do not fall neatly into one specific genre; it is not like discussing, say, game shows: prisons appear across genres and as such seem to have been overlooked' (1995: 186). Furthermore, the subject matter of prisons and media spans several disciplines. Criminologists or media scholars may separately decipher how fictional depictions of crime are presented (Boda and Szabo 2011, Clarke 1990, Kort-Butler and Hartshorn 2011, Mason 2003), or news is reported (Bjornstrom et al. 2010, Schlesinger and Tumber 1994). But, the marriage of the two is relatively unquestioned (Mason 1995: 186).

In the following I look first to discuss the persistence of the spectacle of punishment from public displays to global media such as film, news, but more specifically the television image. I want to consider how this spectacle contributes to a 'politics of amnesia,' which presents criminals in certain ways that can often be concurrent to strategies of penal reform or rehabilitation. In order to ground this discussion, I draw both upon the recent BBC situational comedy (sitcom) *The Visit* (2007) and the Channel 4 documentary film *Sweethearts of the Prison Rodeo* (2009), each indicative of how television as a visual medium allows for certain absences and presences to emerge.

From Old London Bridge to Living Rooms

For the purposes of this chapter, I understand the term spectacle to relate to events that create memorable experiences, particularly with regard to the visual experience. The term has ancient, cultural origins in the Roman staging of circuses and came into Old English through an amalgamation of the French *spectacle* and the Latin *spectaculum*, of 'to view, watch' (Debord 1978). Initial spectacles were freak shows or folk drama, providing wonder and pleasure from strange sights and ridicule (Daston and Park 2001). Certainly in regard to corporeal punishment, the enactment of brutal activities was highly spectacularized. Crowds gathered at sites such as Tyburn Fair or the Old London Bridge to act as both witnesses and guarantors of punishment (Parry 1975; Pratt 2002). However, within Anglo-

American penal systems, the trend towards retribution against the body diminished, and by the twentieth century all penal activities were removed behind the walls of the prison.

Displaced into literature and art, as well as architecture, carceral imaginations were to be stimulated by the emergence of film, presenting to a large and far-reaching audience sights they now rarely saw in everyday life. Since the first prison film in 1919, more than 300 were produced over the course of the twentieth century (Mason 2000b, 33). Prior to the Second World War, it is true to say that few prison films were produced in Britain. In the US, however, films such as *Convict 99* (1938) and *I Am a Fugitive from a Chain Gang* (1932) reflected the contentious issue of corporeal punishment and attempted to provide the public with some kind of insight into prison life, although *Convict 99* was subsequently viewed as too 'soft' a portrayal. However, a period of post-war financial optimism which lasted until the 1960s saw an increase in the number of prisons films being produced; *Good Time Girl* (1949), *The Weak and the Wicked* (1953) and *Now Barabbas* (1948) all highlighted concerns about potential rises in delinquency following a feared 'post-war crime wave'. Prison films of the 1960s such as *The Criminal* (1960), *Birdman of Alcatraz* (1962), and *The Loneliness of the Long Distance Runner* (1962) began, however, to reflect some of the most pressing carceral issues of the decade, including the abolition of hanging in 1965, an expanding prison population and the thorny question of how the penal system could prevent recidivism. Subsequent releases clearly positioned prisons within the social crises and concerns of their time, such as human (prisoner) rights in the 1970s with *Scum* (1979) and a Thatcherite society of polarization and racial inequality in the 1980s with *Burning an Illusion* (1981). As the twentieth-century wore on, prison film (or films with prison scenes) began to fall into more post-modern categories, either reflecting a nostalgia for 'crime free' days or an attempt to tell moral tales, such as *Greenfingers* (2000) and *Tomorrow La Scala* (2002). Although prison films are in no way the largest genre within the film industry, many of them have been popular and critically acclaimed, no doubt because of their ability to open up the prison world to the audience. *The Shawshank Redemption* (1994) was rated number four in *Empire Magazine*'s 'The 500 Greatest Movies of All Time' and number one in the IMDb's Top 250, whilst *The Green Mile* (1999) reached 74th, *The Great Escape* (1963) 109th, and *Cool Hand Luke* (1967) 137th place.

With the displacement of a carceral corporeality, the notion of 'spectacle' has become fundamentally tied up with its technological mediation. A useful commentary on this broad process and its creation of particular forms of subjectivity is provided by Guy Debord, in his influential *The Society of the Spectacle* (1978) (*La Société du Spectacle*). Here, Debord follows up Marx's arguments on alienation by applying commodity fetishism to contemporary mass media. Debord contends that authentic social life, and genuine relations between people, is being replaced by its representation in what he terms the 'degradation of life,' such that, 'all that was once directly lived has become mere representation' (Debord and Sanguinetti 2003: 1). Here, people relate to spectacular events, which

the media represents, and use these to construct their everyday identities, shaping their daily activities and building relationships around commodities. As Debord writes, 'the spectacle is not a collection of images, rather, it is a social relationship between people that is mediated by images' (1978: 4). In the process, popular culture displayed by the mass media creates a sense of 'amnesia' by over-riding genuine events, such as historical episodes that are abhorrent to contemporary ideologies (ibid.: 192). For Debord, radical action was thus required that 'woke up' the spectator 'drugged' by spectacular images, allowing individuals (no longer mere consumers) to reorder their lives, recreating a more self-conscious existence within their world (Ford 2005).

Clearly, the spectacle of the prison persists with the emergence of global media images, and particularly the produced television image, which deploys landscapes of crime, punishment, even simply sublime terrors, as everyday, even banal, spectacles (Garland 1990, Lumby 2002, Nellis and Hayle 1982, Sparks 1992). Thus, the mediated world renders the apparent hidden body of the prisoner much more visible than ever since the removal of public punishment in 1868. In Britain, programmes such as *Dispatches: The Thief Catcher* (2011), *Louix Theroux: Miami Mega-Jail* (2011) and *See Hear: Deaf Prisoners* (2009), alongside documentaries such as ITV's *Holloway* (2009) and *Strangeways* (2011), and the BBC's *Girls Behind Bars* (2011), serve to both illuminate, and in many cases criticize, the 'hidden' world of the penal system, as well as providing for audience entertainment through the presentation of punished bodies as a spectacle. Furthermore, there has also emerged in the UK particularly what I may loosely term the 'prison show' – a wide genre reflecting penchant for situation comedies (sitcoms) such as *Porridge* (1974-1977) *Angry Boys* (2011) and *Let's Go to Prison* (2006); and more hard-hitting drama such as *Prisoner Cell Block H* (1979-1986), *Bad Girls* (1999-2006), *Oz* (1997-2003) and the US show *Prison Break* (2005-2009), which has witnessed its popularity migrate to the UK. Indeed, in 1999 *Bad Girls* achieved an average of 8 million viewers and has arguably 'done more than any other prison drama to advance the cause of penal reform' (Wilson and O'Sullivan 2004: 135). Crime-related television is now estimated to make up 25 per cent of the most popular British television programming (Reiner 2002: 312).

Wilson and O'Sullivan argue that television and film representations are 'an important source of people's implicit and commonsense understandings of prison' (2004: 8) and in the same way have acted as the voice of the prison itself; or, as 'moral fables' (Sparks 1992), warning of the negative aspects of societal transgression. Wilson and O'Sullivan (2004: 35) suggest that since the 1930s, prison films have been indicative of events and background for the time period – in other words, programmes are shaped by the political mood at the time of broadcast. Thus one of the outcomes of (prison) drama is that it 'translates or transcodes ideas and arguments from specialist sources (official reports, social scientific research, experiential writing) into accessible popular forms' (Wilson and O'Sullivan 2004: 14). According to the British Crime Survey (Home Office 2000), 75 per cent of people get their information on the criminal justice system

from the media, but worryingly only 6 per cent consider it inaccurate. Such media representations therefore 'exaggerate the threat of crime and promote policing and punishment as the antidote' (Reiner 2002: 327).

Simultaneously, and likely reactionary to this, there have been marked attempts by prisoners to construct media representations to make their own stories and experience of the prison public. Carnochan (1998) highlights the vast autobiographical literature of prisoners; whereas Churcher (2011) exemplifies inmate-produced media such as The Angolite newspaper, KLSP radio station and LSP-TV television channel based in Louisiana State Penitentiary; which converts simply entertainment through prison television and radio into journalism (ibid.: 388). Similarly using examples from Northeastern USA, Novek contends that these types of media give agency to 'people who are marginalized and despised by society' in order to give transcendence and transformation to the places they inhabit (Novek 2005: 299).

In light of this, it is all the more urgent for scholars working on prisons and carceral spaces more broadly to become attentive to the construction and affective capacity of such imagery. In the next section, I pursue one such line of inquiry, looking to the politics of amnesia.

Spectacle and the Politics of Amnesia

As noted above, media representations are profoundly influential in shaping people's views of prison life. This poses an interesting spatial problem, 'not only because media representations are part of individual and societal conceptions of the world but also because of media's power to conceptualize and spread political ideas and reinforce hegemonic orders' (Zimmerman 2007: 59). In our postmodern world, space-time distanciation between crime and reporting in the media is eroding rapidly (Giddens 1984, Thompson 1995). Thus, 'the present trends indicate a growing symbiosis between media images, criminality, and criminal justice' (Reiner 2002: 328). How, then, do we as social scientists proceed to 'intervene' in this process? In the past, issues of mimeticism, ideological display, or textual deconstruction have been explored (Cresswell and Dixon 2002) – all of which have, arguably, privileged the message rather than the medium – a more recent post-structural attentiveness to ontologies of becoming have pushed geographers particularly to consider how images are constructed with particular affective capacities, such that they 'work' in various ways (Aitken and Dixon 2006). Alongside this, engagements have begun with cultural forms and practices beyond literature and the visual arts that include creative fields such as dance, theatre and music (Cresswell 2006, Leyshon et al. 1998, Malbon 1999, Thrift 1997, 1999). This has paid new attention to our corporeal existence; of particular value here is the observation that visual media is not just optical, but haptical (Crang 2002). How can film, for example 'touch' us, physiologically and emotionally? How is it

that film makes us feel? How does it manifest 'the non-representable that churns [our] stomach or makes [us] smile' (Aitken 2006: 492, Deleuze 1994).

It is at this point, I want to argue, that the term spectacle, as it relates to a political amnesia, can be usefully reworked once more. Previous dialogues relating to amnesia have traditionally concerned themselves with obliterated and 'forgotten' narratives, such as those of colonized indigenous peoples. After the 1960s and decolonization, a new kind of memory discourse emerged which searched for alternative and revisionist histories (Huyssen 1999: 22). National memory, it was argued, had become overly influenced by the global media's focus on certain themes, which contributed towards a politics of amnesia. Thus, 'we do know that the media do not transport public memory innocently. They shape it in their very structure and form' (ibid.: 30). As Rogin suggests, in many cases the cultural form for spectacular displays makes them 'superficial and sensately intensified, short lived and repeatable' (1990, 106), turning domestic citizens into imperial subjects via a forgetting not only of alternate narratives, but of the active construction of narrative forms. Whilst acting in a sometimes voyeuristic way, there is an equal and opposite story to be told about the disadvantages of keeping prisons and spaces of incarceration hidden from public view. Invisibility from the public gaze can render prisoners precarious and exploitable; just has been experienced with media neglect of other marginalized and concealed social groups such as migrant workers and illegal immigrants (Dejanovic 2008, Kihato 2007).

There is considerable scope here to query the way in which media representations provide, as entertainment, a carceral spectacle relating to crime and punishment. But, I want to suggest, there are some subtle twists to how this is accomplished. Spectacle produces a kind of emotion that 'side-steps or forgets (through amnesia) … and is pulled primarily towards the heroism of characters and context' (Aitken and Dixon 2009: 197); and, we can certainly think of viewers as absorbed in the fictional world presented on screen. Yet, television and film as media can also present us with a knowing, ironic, self-referential take on the production of visuals. What this allows for is the destabilizing of spectacle via spectacular means; in other words, media provide for a particular distribution of exposure and denial, reveal and hide.

In order to ground this more subtle play of spectacle I would like to attend to two different examples of carceral media imagery. The first is a BBC sitcom called *The Visit* (2007) and the second a documentary film broadcast by Channel 4 in the UK entitled *Sweethearts of the Prison Rodeo* (2009). Unlike many of the aforementioned programmes which locate characters within prison, *The Visit* is set in a prison visiting room – the only space which the 'outside' visitors are given access to. The visiting room is the interface where we can explore the disjointed lives of the prisoner and their families; a space where different kinds of worlds collide. In many cases this space highlights more profoundly the various repercussions of life in a prison by locating the characters in juxtaposition with their visiting friends and relatives, thus mediating to a degree the relationship between the inside and the outside. Whilst neither examples is totally representative of all types of prison

programme, or indeed the wider cultural phenomenon of media spectacle, they can nevertheless be used to display and discuss how, following the apparently complete incarceration of prisoner bodies, the television media seeks to render this hidden world visible. In this vein, *Sweethearts of the Prison Rodeo,* although different in genre, more practically and less subtly presents the prison (and more specifically the prisoner-body) as a site for spectacle, replicating the fascination for the gallows of old. The analysis that follows is based on deconstructions of both programmes. This involved observation of the context of production, considering the ideologies of and limitations imposed upon the programme-makers, and how these relate to specific neoliberal circumstances in Britain. This alludes to the kind of spectacularization and the themes and issues which the programme makes visible. I have also considered audience feedback where it has been offered on both social networking and television/film review websites, in order to illustrate evidence of affective influence upon viewers.

The Visit

The Visit sets itself in a low-security prison visiting room, and the characters it centralizes are identified as petty crooks, while the friends and family who visit them are portrayed as typical working class citizens. During visiting time prison officers, who offer varying attitudes to the penal system and their role within it, observe them. In *The Visit* there are many occasions where there is the ironic recognition that prison is represented as a spectacle for entertainment purposes. Prisoner Clint's brother, Glen, for instance, who is visiting the prison for the first time, reveals our own feelings as an audience in a comment in episode five: 'Glen: Hiya! Here, it's not like *Prison Break* in 'ere is it?!'[2]

This intertexual reference highlights what Debord (1978) highlights as shaping of our knowledge and assumptions by media consumption. *Prison Break* is a recently popular US prison show that dramatizes the prison as soap-opera, with a multitude of characters, including popular pin-ups, on its main cast list. There is an acknowledgement by *The Visit*'s characters that visitors to the prison would bring with them an image of the penal environment that was created predominantly on the basis of images and representations constructed by the media.

The programme presents a comic stance upon many fundamentally serious issues, such as prisoners struggling to maintain relationships with their friends, wives and children on the outside. And, as a sitcom, and despite their transgression, the audience find the characters likeable, particularly as the absence of any genuinely threatening characters contributes to a lack of representation of the 'ugly' or violent aspects of prison life. This absence can in turn, of course, present a portrayal of prison as 'easy' or 'overly-privileged'. Indeed, for viewers, for most of the series, the view we are presented with most often is a high-angle shot of

2 All direct references to script within *The Visit* and *Sweethearts of the Prison Rodeo* are my personal transcriptions. Therefore any errors or omissions are also my own.

the prison waiting room. This shot, used as both the opening and end credits roll, becomes the emblem and indeed the *only* emblem of prison life – the lack of view of the inner world of the prison environment further contributes to the forgetting of any truly 'hardened' criminals. There is no mention of any violent or sex crimes in *The Visit.* Inmates are portrayed as ordinary people with regular difficulties and problems – the only prison-related problems which the audience are made to consider are those which provide entertainment purposes, particularly Splodge Costello's 'dirty protest' against a lack of toilet roll. Subsequently, audience feedback has been directed towards an ambiguity of the programme. Forum reviews of *The Visit* are unclear as to the allegiance with the genre, with criticism that the 'soft style' nature of the comedy presents unreal characters, which do not comply with the harder messages with which the programme may comply ideologically:

> GedC: Enjoyed this, some great gags and a decent story … the 'safe and gentle' style is becoming the norm in BBC sitcoms … However I think this 'gentle' style is perfect is perfect for *The Visit* because it's not what you expect. What you're expecting from a sitcom set in a prison visiting room is something full of chavs, smack heads and dodgy twats … this show doesn't bother with the obvious, it is played from the point of view of the family. (Forum member of The British Comedy Guide 2007)

Yet, there are many instances where the characters are rendered as 'others'. The visit itself is confined to limited space, which would be similar to the cells in which they spend most of their time. Visitors pass through different levels of security including sniffer dogs and metal detectors. Figure 15.1 illustrates how inmates are also distinguished and divided from visitors by their clothing (a light blue shirt, jeans and standard black trainers), a yellow bib, as well as being made to sit at opposite sides of the table, which are numbered to ensure that each prisoner remains bound to their own assigned territory. These actions promote a particular type of 'forgetting': one which makes the audience (and people more generally) overlook these people as citizens. For Hacking, individuals react to the name or label given to them. He suggests that categorization may 'change the ways in which individuals experience themselves – and may even lead people to evolve their feelings and behaviour' (1999: 104).

And, the same can be observed in the categorisation of prisoners portrayed in *The Visit*. The protagonist Michael claims that his fellow prisoners belong to two categories: 'Smack heads' and 'Knob heads'.

Nevertheless, many viewers perceived the dominant hegemonic message of the text, which was to enjoy *The Visit* as a form of entertainment: 'Mike: Bit delayed I know but I just want to say bravo on an awesome series, best one we have had since *Early Doors* I think … if rubbish like *Tittybangbang* and *Grownups* can get second series then *The Visit* will have no bother' (MySpace user comment 2007).

Figure 15.1 Inmate and visitor – Michael and his Dad
Source: *The Visit* DVD 2007.

These forum activities also illustrate the consumption of a television programme as an active process, whereby people no longer simply accept the dominant messages that are presented by the text. This provides a contrasting opinion to Debord's (1978) contention that there is a lack of active human participation in the creation of meaning.

Aside from the notion that society has forgotten their status as citizens, one of the biggest impacts of incarceration presented to us is the breakdown of relationships. Inmates worry that people on the outside will forget them, or that they will become dispensable, replaceable. With their criminal records, the prisoners are rendered economically undesirable. In social terms, inmates also become expendable. Indeed, we learn that Michael's girlfriend has left him for his best friend. The council has also replaced him with another tenant: his brother Stevie. This being forgotten becomes a long-running joke, as Stevie is now wearing his underpants and his Dad has commandeered his trainers. They seem oblivious to the fact that Michael will need these things, need his life back, once he is released from prison:

> Michael: Have you got my shoes on?!
> Dad: What, yeah. Stevie gave 'em us.
> Stevie: Well they was just lying about your flat, you know?
> Dad: Saved me a couple of quid, eh? Tell you what; they're nipping a bit though.
> Michael: Got me under crackers on an' all?
> [Both laugh]
> Dad: No, Stevie wears them … well, yeah; they're a bit small for me.

Michael: Anything left?
Stevie: Chloe's left.
Michael: I know that.
Stevie: But the good news is she's moved back into your flat.
Michael: You what?
Stevie: Well she's sharing the spare room with Swifty. Dead cosy, or what eh?
Michael: Is that my watch?!
Dad (sheepishly): Well, yeah, it was with your shoes.
Michael: I'm not dead!

Sweethearts of the Prison Rodeo

This spectacular amnesia is also a predominant thread in the documentary film *Sweethearts of the Prison Rodeo.* This documentary details the training and performance of women from the Eddie Warriors Correctional Center at the 2007 Oklahoma State Penitentiary Rodeo held on 17 August – a tradition since 1940 and one of only two rodeos in the world held behind prison walls. Inmates from 12 different prisons across the state compete. As part of the rodeo, inmates participate in dangerous events such as bull riding, team calf roping, bareback riding and the spectacular finale of 'Money the hard way,' where players attempt the perilous task of retrieving a $100 bill attached to a string which is in turn tied to a bull's horn. The female inmates, or 'sweethearts', are very much constructed as heroic characters, brave and determined – exhibiting perfect prisoner behaviour in order to participate. With 30 women trying out for 13 lots, demand is high and the opportunity is viewed as a privilege.

With danger being the pay-off of this privilege, it is not surprising that the dominant rhetoric of this programme provides little comment upon the rodeo as an extravagant, special treatment for prisoners not enjoyed by the 'outside' world. The chosen prisoners are role models for their cohort and even beyond. They are glamorized by the special treatment that being on the rodeo team affords them: whole days spent outdoors, extra exercise, additional recreation and even the heightened sense of comradeship and increased hierarchy that the inmates receive through the wearing of their customized pink rodeo uniform (see Figure 15.2). On the night of the rodeo, the rest of the facility line the route to the transport, cheering and clapping as the rodeo team parade through waving. They are even given the accolade of being 'just as tough as the men'. It is undeniably not uncommon for positive relationships to be constructed in the prison environment – emancipated spaces of incarceration have been documented by the likes of Wilson (2000) and Baer (2005) for example.

However, all of this serves to gloss over the crimes for which they have been committed. Indeed, throughout the narrative such issues are treated with a studied nonchalance. The very name of the programme takes its influence from the wholesome American Country music duo of sisters Janis and Kristine Oliver, entitled *Sweethearts of the Rodeo*. Brandy 'Foxie' Wittle, for example, lists her

Figure 15.2 Female prisoner wearing her pink rodeo uniform
Source: *Sweethearts of the Prison Rodeo* 2009.

felonies which include 10 charges, two drug trafficking offences, two possession of firearms and further charges including possession with intent whilst the camera pans over a backlit sunset view of 'Foxie' on a horse claiming 'I'm not that bad'. The rodeo is presented as an opportunity for change, which further contributes to this amnesic spectacle: (Rodeo captain): 'We've got a lot of girls here who I call role models. They've figured out how to do their time. We can take someone like 'Foxie' here, they've got a career. They've got a lot of confidence now.

The criminality of these women is loosely recorded, with references portrayed as a result of extenuating circumstances such as lack of family ties; Oklahoma's methamphetamine problem; its status as the number one state to lock up women; or merely the naivety of youth. 'Foxie' goes as far as to admit that she 'took the rap' for her boyfriend, while Jamie explains how she got into a bad situation when she was 17, which started as a basic robbery but turned into her shooting a man in the head and her receiving a 30-year sentence. Indeed, the programme goes on to provide a narrative for Jamie as reformed character, centralizing her parole hearing as part of the documentary.

Meanwhile, Danny had so far served 13 years of a 30-year sentence for first-degree murder. Instead of paying attention to this detrimental fact, the Master of Ceremonies at the rodeo instead welcomes this '13-year veteran' to take a bow centre stage while the whole crowd around him cheer. Another inmate admits that he 'consider[s] it an honour to be his friend'. Not only are these prisoners constructed as role models, but they are also winners. 'Foxie' enjoys the dizzy heights of two consecutive years placing first in the bareback riding explaining that the rodeo has become 'the best of times in the worst of times'. As the following

user review illustrates, this alludes to a kind of empathy with the prisoners, which can be contentious for those who confess the need for prison to be a very punitive place:

> It achieves a number of worthwhile things, among them introducing us to criminals so we come to see them as human beings with hopes and dreams in an age where portraying all criminals as monsters is more the rule … Also, some darker aspects of this world are touched on, but not explored. For example, the fact that Oklahoma has almost twice the normal rate of women in prison. (User review of *Sweethearts of the Prison Rodeo* from IMDb 2012)

In documenting this spectacle, however, the film also provides a more critical commentary on spectatorship, as evidenced by the rodeo-goers; the issues raised, it is implied, are just as relevant to the film's TV audience. When asked why they come to watch the rodeo, for the most part people respond that it was an opportunity for families to see their friends and relatives taking part in the deep and fairly prestigious rodeo tradition of Oklahoma State. Yet, for others, the reason is both stark and much darker:

> Rodeo Visitor: It's a gladiator-type thing. We want to come and see people do this who haven't got anything to lose … You'll see people trying to get a hold of something they can't really have and they're still reaching.
>
> Danny: A lot of people just come here to see us get hurt. You know, gladiators, and we go out there to bleed for them.

The audience seem to forget that these are real people. Rodeo audiences in general admit to enjoying the thrills and particularly the spills of these dangerous events, but having prisoners participate gives some an added expectation of those risk-taking non-citizens whose 'lives have less to lose' (Prisoner, *Sweethearts of the Prison Rodeo*). Their status as mothers, fathers, siblings, children, citizens, is forgotten in the narrative of the spectacle of risk and danger: (Danny): 'I've seen guys get messed up pretty bad, you know, it's not just broke[n] arms, broke[n] legs, jaws broke[n]; I mean all kinds of stuff, we're talking about injuries that's gonna be forever.'

The camera technique further spectacularizes these harrowing events for us as TV-viewers. Fast cutting, close range shots of the activities in the bull ring connote chaos and replicate the real-life pace of the action (see Figure 15.3). In other parts of the film, prisoners are frozen mid-air to intensify the force of the bulls' horns and these shots are directly juxtaposed with others of prisoners with blood injuries, facial disfigurements or being attended by medical staff. Some of the most risky activities are even repeated several times throughout the film, much like a goal scored in a football match, inviting the audience to view these occasions as highlights in the documentary.

Figure 15.3 Chaos in the rodeo ring
Source: *Sweethearts of the Prison Rodeo* 2009.

This drive to provide audiences with media coverage of criminal behaviour and its punishment has, of course, also become a key feature of TV news reporting. Recent examples include the live coverage of gunman Raoul Moat's stand-off with police on 9 July 2010 after the shootings of three people including a police officer. News channels broadcast the events 'as they happened' right up until the moment when Moat was shot by police. Barbara Ellen (2010) of *The Observer* wrote that everything that happened '(hiding out, taunting police, blowing your own brains out) could double as a scene from a Bruce Willis action drama'. The television footage, with its carnival atmosphere, leads me to question why it has come to be that the media can now make (tragic) heroes from criminals – a version of reality television that pushed the boundaries of a genre we are becoming more obsessed with (Hughes 2006). As Ellen (2010) notes, 'Moat's sickness met our sickness and we were locked together in a deathly embrace, broken only by adverts'. Clearly a desire to witness punishment has not disappeared with its displacement behind prison walls. Whether it is also a desire for voyeurism or a fascination with the performance of transgression and violence, 'the combination of topicality and limited knowledge of the audience make prisons a suitable subject for the [media]' (Mason 1995: 187).

Conclusion

It is clear that, following the disappearance of corporeal punishment from public space to the 'closed world' of the prison, spectacle of punitive measures has instead been displaced into other cultural constructions such as architecture, literature, and more recently to the visual imagery produced by television and film. Although the spectacle has been well addressed its collaboration with the politics of amnesia make it a key discussion point for scholars interested in emotion and 'affect'. Representations such as TV and film exist as 'more than representations'; they shape popular imaginations and, in turn, have a profound import for policy and intervention. Both *The Visit* and *Sweethearts of the Prison Rodeo* provide important examples of how the television media can spectacularize prison life for a variety of different purposes. In *The Visit*, the absence of any concrete visual representation of the inside of the prison, combined with a creation of likeable characters, performs an amnesiac affect that hides the danger or grim reality of prison life and replaces it with notions of soft prison, or at the very least a nonchalant attitude to the harsh realities of incarceration. In contrast, although *Sweethearts of the Prison Rodeo* paints some of its subjects as heroic pictures, the spectacle of the rodeo environment is made explicit, raising questions around our views of the body of the criminal transgressor as a suitable vessel for brutal, gladiatorial punishments. It is fair to say that the differing attitudes towards the penal system may be attributed to the genre of the programme; however, what is evident is television's powerful capacity, through its constantly-developing visual technologies to generate spectacle that are far more emotive or 'affecting' than anything we have seen before.

In considering the affective nature (in this case the amnesia) of carceral spectacles we can open up the political (prisons and society) at a personal level, learning to critically reflect upon how geographies of inclusion and exclusion, inside and outside, prisoner and civilian, come to be variously ordered and disordered by a contemporary media that fascinates us.

Programmes

In order of appearance in text:
Convict 99 (1938) Gainsborough Pictures
I Am a Fugitive from a Chain Gang (1932) Warner Brothers
Good Time Girl (1949) Sydney Box Productions
The Weak and the Wicked (1953) Marble Arch Productions
Now Barabbas (1948) De Grunwald Productions
The Criminal (1960) Merton Park Studios
The Loneliness of the Long Distance Runner (1962) Woodfall Film Productions
Scum (1979) Berwick Street Productions
Burning an Illusion (1981) British Film Institute

Greenfingers (2000) Boneyard Entertainment
Tomorrow La Scala (2002) BBC
Prison Break (2005-2009) Fox 81 Episodes
The Shawshank Redemption (1994) Castle Rock Entertainment
The Green Mile (1999) Warner Brothers
The Great Escape (1963) The Mirisch Corporation
Cool Hand Luke (1967) Jalem Productions
Dispatches: The Thief Catcher 13 June 2011 C4 20:00
Louix Theroux: Miami Mega-Jail (1) 22 May 2011 BBC2 21:00
Louix Theroux: Miami Mega-Jail (2) 16 July 2011 BBC2 21:00
See Hear: Deaf Prisoners 13 November 2009 BBC1 01:35
Holloway (1) 17 March 2009 ITV1 21:00
Holloway (2) 24 March 2009 ITV1 21:00
Holloway (3) 31 March 2009 ITV1 21:00
Strangeways (1) 9 May 2011 ITV1 21:00
Strangeways (2) 16 May 2011 ITV1 21:00
Strangeways (3) 23 May 2011 ITV1 21:00
Girls Behind Bars (1) 1 March 2011 BBC1 23:20
Girls Behind Bars (2) 8 March 2011 BBC1 23:20
Girls Behind Bars (3) 15 March 2011 BBC1 23:20
Porridge (1974-1977) BBC 20 Episodes
Angry Boys (2011) Australian Broadcasting Corporation 12 Episodes
Let's Go to Prison (2006) Carsey-Werner Company and Strike Entertainment
Prisoner Cell Block H (1979-1986) Reg Grundy and Network Ten 692 Episodes
Bad Girls (1999-2006) ITV 107 Episodes
Oz (1997-2003) HBO 56 Episodes
The Visit (2007) BBC 7 Episodes
Sweethearts of the Prison Rodeo (2009)13 November 2009 C4 22:00

References

Aitken, S.C. 2006. Leading men to violence and creating spaces for their emotions. *Gender, Place and Culture – A Journal of Feminist Geography* 13(5): 491-507.

Aitken, S.C., and D.P. Dixon 2009. Avarice and tenderness in the cinematic: Landscapes of the American West, in *Geography and the Humanities*, edited by M. Dear, D. Richardson and J. Kitchum. London and New York: Routledge, 196-205.

Baer, L.D. 2005. Visual imprints on the prison landscape: A study on the decorations in prison cells. *Tijdschrift Voor Economische En Sociale Geografie* 96(2): 209-217.

Bjornstrom, E.E.S., R.L. Kaufman, R.D. Peterson and M.D. Slater 2010. Race and ethnic representations of lawbreakers and victims in crime news: A national study of television coverage. *Social Problems* 57(2): 269-293.

Boda, Z., and G. Szabo 2011. The media and attitudes towards crime and the justice system: A qualitative approach. *European Journal of Criminology* 8(4): 329-342.

Carnochan, B. 1998. The literature of confinement, in *The Oxford History of the Prison: The Practice of Punishment in Western Society*, edited by N. Morris and D.J. Rothman. Oxford: Oxford University Press, 381-406.

Churcher, K.M.A. 2011. Journalism behind bars: The Louisiana State Penitentiary's Angolite Magazine. *Communication, Culture and Critique* 4(4): 382-400.

Clarke, A. 1990. You're nicked: Television police series and the fictional representation of law and order, in *Come On Down? Popular Media Culture in Post-war Britain*, edited by D. Strinati and S. Wagg. London: Routledge, 232-53.

Clarke, K. 2011. Punish the feral rioters, but address our social deficit too. *The Guardian* [Online]. Available at http://www.guardian.co.uk/commentisfree/2011/sep/05/punishment-rioters-help [accessed 5 September 2011].

Crang, M. 2002. Rethinking the observer: Film, mobility and the construction of the observer, in *Engaging Film: Geographies of Mobility and Identity*, edited by T. Cresswell and D.P. Dixon. Lanham, MD: Rowman and Littlefield, 13-31.

Cresswell, T. 2006. *On the Move: Mobility in the Modern Western World*. London: Routledge.

Cresswell, T., and D.P. Dixon (eds) 2002. *Engaging Film: Geographies of Mobility and Identity*. Lanham, MD: Rowman and Littlefield.

Daston, L., and K. Park 2001. *Wonders and the Order of Nature 1150-1750*. New York: Zone Books.

Debord, G. 1978. *Society of the Spectacle and Other Films*. London: Rebel.

Debord, G., and G. Sanguinetti 2003. *La Véritable scission dans l'internationale*, in English: The Real Split in the International. London: Pluto Press.

Dejanovic, S. 2008. Invisible bodies, illusionary securing: The performance of illegality at the US-Mexico border, in *Violent Interventions: Selected Proceedings of the Fifteenth Annual Conference of the York Centre for International and Security Studies*, edited by M. Ayyash and C. Hendershot. Ontario: York Centre for International and Security Studies, York University, 81-104.

Deleuze, G. 1994. *Cinema I: The Movement-image*. London: Athlone Press.

Ellen, B. 2010. Now Raoul Moat is dead, perhaps we should all feel a little sick. *The Observer* [Online]. Available at http://www.guardian.co.uk/commentisfree/2010/jul/11/raoul-moat-rape-ronald [accessed 24 September 2011].

Ford, S. 2005. *The Situationist International: A User's Guide*. London: Black Dog Publishing.

Foucault, M. 1977/1991. *Discipline and Punish: The Birth of the Prison*. London: Penguin.

Garland, D. 1990. Frameworks of inquiry in the sociology of punishment. *The British Journal of Sociology* 41(1): 1-15.

Giddens, A. 1984. *The Constitution of Society: Outline of the Theory of Structuration*. Cambridge: Polity Press.

Gilmore, R.W. 1999. Globalisation and US prison growth: From military Keynesianism to post-Keynesian militarism. *Race and Class* 40(2-3): 171-188.

Gilmore, R.W. 2007. *Golden Gulag: Prisons, Surpluses, Crisis, and Opposition in Globalizing California*. Berkeley: University of California Press.

Graduate School of Architecture Planning and Preservation of Columbia University New York. 2006. Million Dollar Blocks. *Spatial Information Design Lab* [Online]. Available at http://www.spatialinformationdesignlab.org/projects.php?id=16 [accessed 2 September 2009].

Gregory, D. 2006. The black flag: Guantanamo Bay and the space of exception. *Geografiska Annaler Series B-Human Geography* 88B(4): 405-427.

Gregory, D. 2007. Vanishing points: Law, violence and exception in the global war prison, in *Violent Geographies: Fear, Terror and Political Violence*, edited by D. Gregory and A. Pred. New York: Routledge, 205-236.

Hacking, I. 1999. *The Social Construction of What?* Harvard: First Harvard University Press.

Home Office. 2000. British Crime Survey. *National Archives* [Online]. Available at http://webarchive.nationalarchives.gov.uk/20100413151441/http://www.homeoffice.gov.uk/rds/pdfs/hosb1800.pdf [accessed 5 October 2011].

Horschelmann, K. 1997. Watching the East – Constructions of 'otherness' in TV representations of East Germany. *Applied Geography* 17(4): 385-396.

Hughes, R. 2006. TV genre popularity based on profit. *The Daily Helmsman 17 March 2006* [Online]. Available at http://media.www.dailyhelmsman.com/media/storage/paper875/news/2006/03/17/News/Tv.Genre.Popularity.Based.On.Profit-1757139.shtml [accessed 28 September 2011].

Huyssen, A. 1999. Present pasts: Media, politics, amnesia. *Public Culture* 12(1): 21-38.

International Movie Database (IMDb). 2012. Reviews and ratings for *Sweethearts of the Prison Rodeo*. *IMDb* [Online]. Available at http://www.imdb.com/title/tt1095414/reviews [accessed 14 April 2012].

Johnson, B.R. 2004. Religious programs and recidivism among former inmates in prison fellowship programs: A long-term follow-up study. *Justice Quarterly* 21(2): 329-354.

Kihato, C.W. 2007. Invisible lives, inaudible voices? The social conditions of migrant women in Johannesburg. *African Identities* 5(1): 89-110.

Kort-Butler, L.A. and K.J.S. Hartshorn 2011. Watching the detectives: Crime programming, fear of crime, and attitudes about the criminal justice system. *Sociological Quarterly* 52(1): 36-55.

Leyshon, A., D. Matless and G. Revill (eds) 1998. *The Place of Music*. New York: Guilford Press.

Lumby, C. 2002. Televising the invisible: Prisoners, prison reform and the media, in *Prisoners As Citizens: Human Rights in Australian Prisons*, edited by D. Brown and M. Wilkie. Sydney: Federation Press, 103-112.

Malbon, B. 1999. *Clubbing: Dancing, Ecstacy, Vitality*. London: Routledge.

Mason, P. 1995. Prime time punishment: The British prison and television, in *Crime and the Media: The Post-Modern Spectacle*, edited by D. Kidd-Hewitt and R. Osbourne. London: Pluto, 185-205.

Mason, P. 2000a. Lies, distortion and what doesn't work: Monitoring prison stories in the British media. *Crime Media Culture* 2(3): 251-267.

Mason, P. 2000b. Watching the invisible: Televisual portrayal of the British prison 1980-1990. *International Journal of the Sociology of Law* 28: 33-44.

Mason, P. 2003. *Criminal Visions: Media Representations of Crime and Justice*. Portland, OR: Willan Publishing.

MySpace. 2007. The Visit. *MySpace* [Online]. Available at http://www.myspace.com/the_visit [accessed 21 September 2010].

Nellis, M., and C. Hayle (eds) 1982. *The Prison Film*. London: Radical Alternatives To Prison.

Novek, E.M. 2005. 'Heaven, hell, and here': Understanding the impact of incarceration through a prison newspaper. *Critical Studies in Media Communication* 22(4): 281-301.

Parry, L.A. 1975. *The History of Torture in England*. Montclair, NJ: Patterson Smith Publishing Corporation.

Pratt, J. 2002. *Punishment and Civilization*. London: Sage.

Reiner, R. 2002. Media made criminality: The representation of crime in the mass media, in *The Oxford Handbook of Criminology*, edited by M. Maguire, R. Morgan and R. Reiner. Oxford: Oxford University Press, 302-341.

Rogin, M. 1990. Make My Day! Spectacle as amnesia in imperial politics. *Representations* 29: 99-123.

Schlesinger, P., and H. Tumber 1994. *Reporting Crime: The Media Politics of Criminal Justice*. Oxford: Clarendon Press.

Sparks, R. 1992. *Television and the Drama of Crime: Moral Tales and the Place of Crime in Public Life*. Buckingham: Open University Press.

Spergel, I.A. 1990. Youth gangs: Continuity and change, in *Crime and Justice: An Annual Review of Research*, vol. 12, edited by N. Morris and M. Tonry. Chicago, IL: University of Chicago Press, 171-275.

The British Comedy Guide. 2009. The Visit – forum thread. *The British Comedy Guide* [Online]. Available at http://www.comedy.co.uk/forums/thread/7993/ [accessed 23 September 2009].

Thompson, J. 1995. *The Media and Modernity: A Social Theory of the Media*. Oxford: Polity Press.

Thrift, N. 1997. The still point: Resistance, expressive embodiment and dance, in *Geographies of Resistance*, edited by S. Pile and M. Keith. London: Routledge, 124-151.

Thrift, N. 1999. Steps to an ecology of place, in *Human Geography Today*, edited by D. Massey. Cambridge: Polity Press, 295-322.

Turner, J. (forthcoming). Displacing landscapes of incarceration: Spectacle, crime and punishment in the TV sitcom. *Aether: The Journal of Media Geography*.

Wilson, A. 2000. There is no escape from Third-Space Theory: Borderland discourse and the in-between literacies of prison, in *Situated Literacies: Reading and Writing in Context*, edited by D. Barton, M. Hamilton and R. Ivanic. London; New York: Routledge, 54-69.

Wilson, D., and S. O'Sullivan 2004. *Images of Incarceration Representations of Prison in Film and Television Drama*. Winchester, UK: Waterside Press.

Zimmerman, S. 2007. Media geographies: Always part of the game. *Aether* 1: 59-62.

Chapter 16
Dialogues across Carceral Space: Migration, Mobility, Space and Agency

Nick Gill, Deirdre Conlon and Dominique Moran

We can identify various sets of existing debates about incarceration and confinement that the volume contributes towards. In what follows we group the contributions into four areas that point to innovative engagements in critical human geography where prison studies, penology and migration studies inform new understandings of carceral spaces. First, while much has already been written about the role of space in carceral settings, the chapters included here provide a unique and original angle onto these debates. Second, the chapters signal the importance of attending to a range of mobilities in the context of prison and migrant detention. While this may seem counter-intuitive because these sites seek, first and foremost, to immobilize those they contain, we argue that perspectives that can account for the mobility of various elements of the carceral environment promise to enrich our understanding of confinement in important new ways. Third, the contributions have raised the issue of the degree to which resistance is possible and desirable in carceral environments in new and innovative ways. And fourth, the volume has highlighted the importance of attending to immigration detention as it relates to both prison and migration, as well as integrating each of these into a holistic carceral framework.

The volume has highlighted various issues surrounding the role of space in penal settings. Much writing about incarceration and prisons is concerned with whether the border between prisons and non-prisons is more permeable than we tend to assume (see, for example Baer and Ravneberg 2008). Debate has centred on the imagination of the so-called 'total institution' (Goffman 1961), which sees prisons as separated from society not just by physical borders and barriers but also in terms of the thickness and density of institutionalized practices and routines. The contributions in this volume point to a variety of spatialized considerations that promise to enrich this debate. The contributions from Moran et al, De Dardel, McWatters and Michalon all question the supposedly clear-cut distinction between prison and non-prison space. They draw attention to a liminal, in-between space that can act to blur the distinction between prison and non-prison space, such as in the case of visiting rooms and 'open' prisons. Here we witness an interpenetration of worlds rather than a demarcation and separation. And it is here, at this confluence, that we would locate carceral spaces rather than simply prison spaces. The former includes not only traditional prisons but all the grades and varieties

of confinement that are possible outside formal prison systems. This includes spaces of confinement that fall under civil law, such as immigration detention, as well as the forms of confinement that burst internment structures and deliver carceral effects without physical immobilization, such as electronic monitoring, surveillance and securitized public spaces.

A second debate surrounding penal spaces and penology more broadly, which is taken up in Gill's chapter, concerns the relations of dependence and production between prison and non-prison spaces (see also Wacquant 2001, Peck 2003, Bonds 2009). Jean Baudrillard diagnoses the contemporary human condition as one in which the boundary between the real and the virtual has been radically blurred. Much of life is now lived through virtual encounters, through simulations, to the extent that not only is the 'real' eclipsed, but it is, in fact, lost to the extent that there exists no independent and traceable real 'beneath' or 'behind' the simulations. In an almost throw-away note about prisons and contemporary society, Baudrillard brilliantly describes one of the functions of carceral spaces in these terms. In the course of developing an example about Disneyland, which Baudrillard sees as functioning in order to make us believe that 'the rest is real' (1994: 12) (when all the time the America that surrounds it is just as much a fantasy and simulation as Disneyland itself) Baudrillard comments that carceral space operates in much the same way. As he writes, 'prisons are there to hide that it is the social in its entirety, in its banal omnipresence, that is carceral' (ibid.: 12), a point that is alluded to in Turner's discussion in this volume of televisual carceral simulation.

The interaction of virtual and actual carceral environments, which produces a circuit of prison and prisoner representations, appears to support Baudrillard's observations. These representations exist in their own right but are also performative in content, reflecting and contributing towards the atmosphere and culture of carceral settings. Following Baudrillard's insights, it is through these mechanisms of representation – such as the prisons that exist in written and televisual form as discussed by McWatters and Turner in their chapters (see also Carnochan 1998) – that the claim of the separateness of prison from everyday 'civilian' life is most clearly articulated. And yet the contributions in this volume draw attention to the various linkages and symbiotic relations between the imagined inside and outside of prisons and detention centres that betray the essential, and increasingly clear, continuity between them.

This volume has underscored the rich symbolic, economic, political and social ways in which prisons and non-prisons share such 'symbiotic' relations (Wacquant 2001). Most obviously, opportunities for virtual encounters blur the imagined prison/non-prison boundary through calls, videolinks and recordings that can deliver aspects of the presence of the incarcerated into a range of situations including the lives of their loved ones and the courts that consider their claims. More broadly, Nowakowski's chapter in this volume underscores the economic interdependencies between prison and non-prison spaces in terms of local labour markets and their embeddedness into international circuits of capital that often search out exploitable low wage labour reserves. The circulation of capital and

labour through prison economies, which offer cheap sources of labour, strong motivations for relocation, and both promises and threats to the local economies that house them, constitute a key consideration for understanding carceral spaces (Gilmore 2007). The political disjunction between prison and non-prison spaces as discussed in Mitchelson's chapter, including the dividend that mainstream politicians and political parties can derive from the careful placing and regulating of formal political (non)participation of inmates, concomitantly underscores the political contingencies that traverse the supposed prison/non-prison divide.

De Dardel's chapter raises a further issue concerning the relation between carceral environments and space. The internationalizing nature of the systems of prison and detention can make policy transfer easier as well as rendering prison personnel less accountable. Internationalization of carceral systems can also make invisibility and remoteness of inmates and immigration detainees easier to achieve (Mountz 2010). Such internationalization often goes hand in hand with privatization, as global corporations take on the day to day running of many carceral establishments. The circulation of expertise and personnel, such as prison officials, who often display high rates of staff turnover due to stressful conditions, as well as elite figures including judges and institutional directors who orbit within a trans-carceral space that integrates various degrees and types of incarceration into a specific governmental rationality, therefore becomes easier. Attending to the ways in which policies and practices transplant from one carceral space to another thus emerges as crucial. Policy transfer can be across relatively localized national contexts, across different carceral tracks (such as immigration and criminal, or jail and prison in the US) or across diverse international contexts. Key questions that need to be answered in terms of policy diffusion concern the degree to which the law leads or trails practice in situ (Weber 2003), the speed of policy migration (Peck 2002), the degree to which policies 'transfer' from one site or country to another or travel in more complex and emergent ways (Prince 2010) and the degree to which practices and policies are determined top-down or bottom-up through various points of pressure, including (the threat of) media attention.

Alongside highlighting a set of important debates about space, another of the volume's themes has emphasized the importance of attending to mobility within and beyond carceral environments. We see this as a particularly rich lens though which to apprehend current developments in carceral spaces. The circuitry of the incarcerated, between institutions as well as into other contexts of confinement, forms a primary concern. The frequent spatial relocation of inmates and migrant detainees for example, as discussed by Michalon and Hiemstra in this volume, is a striking characteristic of carceral spaces (see also Moran et al. 2012, Gill 2009, 2014). As Hiemstra discusses in her chapter, movements and transfers can be as a result of inmates' and detainees' processing or their status within their respective carceral systems, and can be justified on the basis of cost, security, sentence severity, informal punishment and efficiency. In any event it serves a centrally important function in shoring up and performing the symbolic power of the prison, especially in terms of induction and initiation. As Moran et al.'s

chapter emphasizes, the combination of long-distance movement and entry into the prison system can have particularly powerful psychological and emotional effects over inmates.

It is not just inmate and detainee mobility that is highlighted by a focus on mobility within prisons however. One can also think of the mobility of personnel, elite decision makers, economic resources, policies and practices, loved ones and material objects throughout and alongside carceral systems. Such an analytical perspective moves beyond place-based views of prison and offers the tools to understand how prisons are situated within shifting and dynamic networks of interactions.

A further debate concerns the relationship between carceral spaces/ incarceration and detention and agentic resistance. De Dardel's chapter asserts the impossibility of foreclosing everyday resistance within prisons (see also Feldman 1991). She nevertheless advocates a nuanced view of the notion of resistance, and Conlon's chapter encourages an approach that avoids the pitfalls of becoming locked into a battle 'against' hegemonic configurations that is ultimately contained within the same system of power that it seeks to counter. Approaches that are beyond counter-hegemonic (Day 2004) seek to sidestep dominant tropes rather than reproduce them in opposition, and while this can be challenging there are spaces and times within carceral environments when this becomes possible.

One specific further contribution of the volume is to develop our understandings of the relationship between immigration detention and broader carceral spaces and practices. Among scholars who examine immigration detention there is concern over the privatization of detention spaces and the difficulties of organizing and resisting immigration detention. This is compounded, as Martin outlines in her chapter, by the increasing discretion among front-line asylum and migrant support sector personnel that can undermine the accountability of the overall system and contribute to the acute disorientation that immigration detainees experience within carceral networks. Both of these issues resonate with developments in 'mainstream' prison settings, but take on different implications and connotations in an immigration context.

Hence the interface of mainstream prison and immigration detention is centrally important. On the one hand the two carceral systems are often imagined to be distinct, and this claim crystallizes into differences in accountability and provision on both sides of this divide. On the other hand, however, the distinction becomes more dubious where detainees experience prison conditions, are housed in prisons, and where they are increasingly the subject of discourse and practices incepted with criminals in mind, as both Hiemstra and Martin discuss in their chapters (see also Bosworth and Guild 2008). The interpenetration of these internment systems in particular creates opportunities for experimentation, transplantation of policies and practices, and streamlining strategies that display a complex and vicarious politics. The points at which the mainstream and detention systems merge and diverge must therefore constitute a particular area of interest not just for this volume but more broadly for students of carceral space.

This is not to say that we feel the contributions presented here exhaust the possible research avenues available to a closer engagement between carceral spaces and geographical perspectives and approaches. Prominent among the issues that deserve closer attention in future work are the racialized character of many prison practices (Gilmore 2007, Wacquant 2001), the impact of prison and other forms of confinement on partners, children and other loved ones (Harman et al. 2007), the emotional complexities of prison and detention spaces, especially in the context of what Martin's chapter identifies as the increasing agency and discretion of employees within carceral institutions (Hall 2012, Weber 2003), and the management of the micro-spaces of juridico-political regulation, both front-stage and back-stage (Rock 1993).

A further theme that we feel is important concerns the politics of carceral space. There are concerns among geographers and social scientists more broadly that increasing numbers of individuals and sites of everyday life are situated beyond politics and that an increasing number of political decisions are foreclosed, reducing political contestation to a consideration not of genuine alternatives but to technical deliberations about how best to achieve ends that are already settled and unchallenged (Swyngedouw 2010, Mouffe 2005). In a punitive context such settlements might include the rise of pre-emptive sanctions and the systematic abandonment of genuine commitment to rehabilitation among many developed country juridico-political systems (Anderson 2010, Rose 2000). Such 'post-political' devices (Žižek 1999) act to shut down dissenting voices and the contributions to this volume hint at at least two ways in which this is achieved in prisons. The first is through the overt design of space in order to secure control. Prisons are designed in order to contain various things as Milhaud and Moran's chapter discusses, including disease, fire and, perhaps most importantly, unrest. The architectural expression of the fear of contagion and the institutionalized determination to contain it reaches its zenith in prison buildings such that the containment of political disruption and dissent is an explicit aim of prison environments. In this sense the political is strictly harnessed in these settings through the appropriation of space.

Relatedly, the imagined circuits of crime and behaviours that are taken to lead to recidivism within carceral spaces are important in determining policies and practices, even if these are based upon little more than suspicion or fear of possible futures. In this sense the imagined circulation of contagion of behaviours and attitudes structures prison spaces at least as much as contagion per se, performing a self-fulfilling virtualism in the governance of carceral environments (Carrier and Miller 1998).

Second, in Žižek's analysis the emphasis is placed upon how post-political mechanisms often do not over-rule or prohibit political claim making, but absorb claims within a technical framework that prevents individual grievances ever gaining enough momentum to produce group disenfranchisement. In this way, post-political mechanisms are intensely reactive to this or that claim, with the effect that specific claims never become universalized. And it is undeniable that

immigration detention spaces, for example, are adept at directing political energies towards technical issues (see Mountz 2010). There are mechanisms for the questioning and appeal of conditions, food, visitor access and general detention centre administration, which operate more or less effectively, all the while distracting from the fact that immigration detainees are incarcerated without trial, often for indefinite periods of time and in conditions that would embarrass even mainstream prison managers (see Dikeç 2005). This dissipation and refraction of political critique we feel is an important issue for future research. Conlon's chapter in this volume highlights the difficulties and complexities of dissent within carceral environments and underscores the importance of theorizing this area further.

From a different direction, the siting and emplacement of prison and detention establishments themselves, the locations of which are negotiated over, desired, feared and contested, constitutes another thread in the construction and maintenance of carceral spaces and deserves closer scrutiny (Bonds 2009). Establishments display a lifecycle of use and decline and while they are eventually destroyed at least in physical form, their sites are often haunted by traces of their presence long after their formal dismantlement (Graham and McDowell 2007). Attending to these lifecycles of physical space will hold lessons for how prison establishments can be planned and integrated into existing landscapes in the future.

What is more, we hope that the volume has paved the way for conceptual work that elucidates and marks out emerging spaces of carceral activity: the distinctions between incarceration, constraint, prison, confinement, internment, immobility and a lack of liberty need to be carefully theorized in order to develop rich frameworks and lexicons for apprehending the evolving character of carceral spaces. And, aside from mobility within carceral settings as a key conceptual lens through which to understand – and indeed keep pace with – contemporary innovations in confinement, the language of borders, liminality, interdependence, contingency and symbiosis promise to offer rich theoretical distinctions and perspectives onto the carceral landscape beyond the prison/non-prison binary.

Finally we note that much of the work within carceral geography originates in or pertains to the highly incarcerative, or 'hypercarcerative' contexts of the US, the UK and the Russian Federation, raising questions over the transferability of theorizations of the carceral to other less carcerative, or actively 'de-carcerative' settings. We would therefore advocate closer links between comparative criminal justice (a long-standing sub-discipline within criminology) and place-based comparisons of different carceral systems, as well as greater attention to how these systems might be changed. Carceral geography (see Moran 2013, Philo 2012) promises to provide this spatial comparative perspective and, as it gains momentum, should seek to foreground non-Western voices and perspectives (especially those of the confined themselves) where possible.

In general, it is clear that confinement in its broadest sense has over-spilt formal prisons. Constraints over human movement are increasingly discernible around borders (which are themselves spatially proliferating, see Vaughan-Williams 2010), gated communities, elite or exclusive 'public' spaces and events (Mitchell

2003), privatized and cordoned rural spaces and transport nodes and linkages (Martin 2010). Moreover, the fact that many countries have reached the capacity of their formal stock of prisons has resulted in the production of systematic linkages between areas of the city, social classes and races and carceral systems (Wacquant 2001, Gilmore 2007, Bosworth 2010), as well as producing a range of innovations that extend carceral practices well beyond prison walls.

This book demonstrates both that geographers and geography have much to contribute to debates surrounding incarceration and that there is something distinctive about geography that can create perspectives onto carceral spaces that are unique, timely and hugely consequential for other disciplinary perspectives as well as for policymakers in this field. The contributions have illustrated that the emerging field of carceral geographies is developing in important and productive directions and beginning to engage with a range of theoretical issues, beyond the usual suspects. They have exemplified an international perspective that attends on the one hand to multiple country locations of incarceration and confinement, and on the other to the increasing internationalization, inter-connection and transnationalism of carceral systems. They have illustrated the importance, and affirmed the possibility, of using a variety of methodological approaches – from interviews, focus groups and surveys to visual analysis and cartography – in order to apprehend developments in contemporary carceral conditions. And most fundamentally, they have signalled the importance of thinking about carceral environments spatially, in order to understand them and, from there, in order to build informed responses to the ways in which they are currently evolving.

References

Anderson, B. 2010. Preemption, precaution, preparedness: Anticipatory action and future geographies. *Progress in Human Geography* 34: 777-798.

Baer, L. and B. Ravneburg 2008. The outside and inside in Norwegian and English prisons *Geografiska Annaler B* 90(2): 205-216.

Baudrillard, J. 1994. *Simulacra and Simulation*, University of Michigan Press

Bonds, A. 2009. Discipline and devolution: Constructions of poverty, race and criminality in the politics of rural prison development, *Antipode* 41(3): 416-438.

Bosworth, M. 2010. *Explaining U.S. Imprisonment*. London: Sage.

Bosworth M. and M. Guild 2008. Governing through migration: Security and citizenship in Britain. *British Journal of Criminology* 48: 703-719.

Carnochan, B.W. 1998. The literature of confinement, in *The Oxford History of the Prison: The Practice of Punishment in Western Society*, edited by N. Morris and D.J. Rothman. Oxford: Oxford University Press, 380-406.

Carrier, J. and D. Miller 1998. *Virtualism: A New Political Economy*. Oxford: Berg.

Day R. 2004. From hegemony to affinity. *Cultural Studies* 18: 715-748.

Dikeç, M. 2005. Space, politics, and the political, *Environment and Planning D: Society and Space* 23(2): 171-188.

Feldman, A. 1991. *Formations of Violence: The Narrative of the Body and Political Terror in Northern Ireland.* Chicago and London: The University of Chicago Press.

Gill, N. 2009. Governmental mobility: the power effects of the movement of detained asylum seekers around Britain's detention estate. *Political Geography* 28: 186-196.

Gill, N. 2014. *Peopling Immigration Control: Geographies of Governing and Activism in the British Asylum System*, Oxford: Wiley-Blackwell.

Gilmore, R.W. 2007. *Golden Gulag: Prisons, Surplus, Crisis, and Opposition in Globalizing California*. Berkeley: University of California Press.

Goffman, E. 1961. *Asylum: Essays on the Social Situation of Mental Patients and Other Inmates*, Garden City, NY: Anchor Books.

Graham, G. and S. McDowell 2007. Meaning in the Maze: The heritage of Long Kesh. *Cultural Geographies* 14(3): 343-368.

Hall, A. 2010. 'These People Could Be Anyone': Fear, contempt (and empathy) in a British immigration removal centre. *Journal of Ethnic and Migration Studies* 36: 881-898.

Harman, J.J., V.E. Smith and L.C. Egan 2007. the impact of incarceration on intimate relationships. *Criminal Justice and Behaviour* 34: 794-815.

Kofman, E. 2004. Family-related migration: Critical review of European studies. *Journal of Ethnic and Migration Studies* 30(2): 243-263.

Latour, B. 2010. *The Making of Law: An Ethnography of the Conseil d'Etat.* Cambridge: Polity Press

Martin, L. 2010. Bombs, bodies, and biopolitics: Securitizing the subject at the airport security checkpoint. *Social & Cultural Geography* 11(1): 17-34.

Mitchell, D. 2003. *The Right to the City: Social Justice and the Fight for Public Space*. New York: Guilford Press.

Moran, D. 2014 (forthcoming), *Carceral Geography: Prisons, Power and Space.* Farnham: Ashgate.

Moran, D, L. Piacentini and J. Pallot 2012. Disciplined mobility and carceral geography: Prisoner transport in Russia. *Transactions of the Institute of British Geographers* 37: 446-460.

Mouffe, C. 2005. *On the Political.* Abingdon: Routledge.

Mountz, A. 2010. *Seeking Asylum: Human Smuggling and Bureaucracy at the Border*. Minneapolis: University of Minnesota Press.

Ogborn, M. 1995. Discipline, government and law: Separate confinement in the prisons of England and Wales, 1830-1877. *Transactions of the Institute of British Geographers* 20: 295-311.

Peck, J. 2002. Political Economies of Scale: Fast Policy, Interscalar Relations, and Neoliberal Workfare. *Economic Geography* 78: 331–360.

Peck, J. 2003. Geography and public policy: Mapping the penal state. *Progress in Human Geography* 27(2): 222-232.

Philo, C. 1989. 'Enough to Drive One Mad': The organisation of space in nineteenth century lunatic asylums, in *The Power of Geography: How Territory Shapes Social Life*, edited by J. Wolch and M. Dear. London: Unwin Hyman.

Philo, C. 2012. Security of geography/geography of security. *Transactions of the Institute of British Geographers* 37: 1-7.

Prince, R. 2010. Policy transfer as policy assemblage: Making policy for the creative industries in New Zealand. *Environment and Planning A* 42(1): 169-186.

Rock, P. 1993. *The Social World of an English Crown Court: Witness and Professionals in the Crown Court Centre at Wood Green*. Oxford and New York: Clarendon Press.

Rose, N. 2000. Government and control. *British Journal of Criminology* 40(2): 321-339.

Swyngedouw, E. 2010. Apocalypse forever? Post-political populism and the spectre of climate change. *Theory Culture & Society* 27(2-3): 213-232.

Vaughan-Williams, N. 2010. The UK border security continuum: Virtual biopolitics and the simulation of the sovereign ban, *Environment and Planning D: Society and Space* 28: 1071-1083.

Wacquant, L. 2001. Deadly symbiosis: When ghetto and prison meet and mesh. *Punishment & Society* 3(1): 95-133.

Weber, L. 2003. Down that wrong road: Discretion in decisions to detain asylum seekers arriving at UK ports. *The Howard Journal* 42: 248–262.

Žižek, S. 1999. *The Ticklish Subject: The Absent Centre of Political Ontology*. London: Verso.

Index